SURPRISED
by
SCRIPTURE

SURPRISED

by

SCRIPTURE

Engaging Contemporary Issues

N. T. WRIGHT

HarperOne
An Imprint of HarperCollinsPublishers

HarperOne

SURPRISED BY SCRIPTURE: *Engaging Contemporary Issues*. Copyright © 2014 by N. T. Wright. All rights reserved. Printed in the United States of America. No part of this book may be used or reproduced in any manner whatsoever without written permission except in the case of brief quotations embodied in critical articles and reviews. For information address HarperCollins Publishers, 195 Broadway, New York, NY 10007.

HarperCollins books may be purchased for educational, business, or sales promotional use. For information please e-mail the Special Markets Department at SPsales@harper collins.com.

HarperCollins website: http://www.harpercollins.com

HarperCollins®, ◾®, and HarperOne™ are trademarks of HarperCollins Publishers.

FIRST EDITION

Library of Congress Cataloging-in-Publication Data is available upon request.

ISBN 978–0–06–223053–9

14 15 16 17 18 RRD(H) 10 9 8 7 6 5 4 3 2 1

For Francis Collins

CONTENTS

PREFACE

SEVERAL OF THESE essays have been as much a surprise to their author as they may be to their readers. I have spent much of my life trying to understand the New Testament in its own context, and so have often had to postpone consideration of other subjects, however fascinating in themselves. However, since I am also committed to relating the Bible to our own day, some of these other subjects have forced themselves upon my attention from time to time. The present collection of pieces is the result. I offer them, for what they may be worth, not as dogmatic pronouncements but as explorations into vast and exciting topics.

In almost all cases, I did not take the initiative to write on these subjects. I was responding, as best I could, to questions that others had raised, and to the frequent invitation to explore contemporary issues from a biblical perspective. I am grateful for these requests and for the many exciting discussions into which they have led me.

I have smoothed out some of the original lecture format, but I have not attempted to turn these originally independent pieces into a single exposition. My hope has been that each piece will stand on its own merit. For this, I have often had to restate certain basic points which are common to several of the subjects. The reader who works through the pieces one by one will therefore encounter a certain amount of repetition, for which I apologize.

I suspect, then, that many will indeed be surprised by the results. It is widely assumed, for instance, that the Bible offers a particular view on topics like science and religion, the ordination of women, and Christian involvement in political life—and in each case the view in question is not the one to which the texts have led me. The "surprise" in the title thus refers both to the fact that many people may not expect the Bible to have much to say on these topics in the first place, and also, second, to the fact that when it does speak to them it may not say what people have imagined.

When I stand back from these essays, various patterns and themes emerge. First, I am frequently addressing an American context, though I hope what I say is more broadly relevant as well (not least, of course, because of the powerful influence of American culture in the rest of the world). The reason for this focus is obvious: I have frequently been invited to lecture in the United States, and most of these papers were originally given there. I am well aware of the dangers of trying to assess cultures other than my own on the basis of small acquaintance. However, since I have had the privilege of living in America for two extended periods, as well as making a large number of shorter visits, I hope I have not made too many obvious mistakes. There are times when it is good "to see ourselves as others see us," and I hope these reflections from a friendly outsider may prove helpful.

Second, one of the features of the contemporary Western world to which I draw attention repeatedly in this book is the massive influence, often unrecognized, of the particular philosophy we associate with the eighteenth-century Enlightenment. I have come to the view that, unless we glimpse the roots of what today is taken for granted in our world, we will not understand why we see problems the way we do, and will not appreciate what the Bible might have to say about them. When people think of "living in the modern world," very often what they are doing is embracing one particular ancient philosophy (Epicureanism) in a modern guise. As C. S. Lewis regularly remarked, the chronological snobbery of the modern age (i.e. the assumption that anything that comes after around 1750 is somehow superior to anything that went before) needs confronting at several levels.

Third, since the book is drawing on various parts of the Bible to address contemporary issues, it is inevitable that some of the material overlaps with biblical exegesis I have offered elsewhere, for instance in my series of small commentaries (*Matthew for Everyone,* and so on, published by SPCK in London and Westminster John Knox in Louisville), and also in books like *Surprised by Hope* (SPCK and HarperOne) and *Evil and the Justice of God* (SPCK and IVP). Those familiar with these books will recognize some of the material. I hope that others will be drawn by these essays into those wider discussions.

The original setting of each of the essays is noted in the Afterword. I am very grateful to the many organizations and individuals who pressed me to come and talk to them on these subjects, and whose welcome, friendship, and (not least) questions and challenges have been important at several levels.

One friend has been a particular inspiration, and this book is dedicated to him. Professor Francis Collins has played a leading and courageous role in bridging one of the great divides in contemporary American culture, that between science and faith. Our shared love of music has provided a delightful third element to our friendship: testimony, perhaps, to the fact that some of the most important questions in life need to be approached from several different angles at once. My warm thanks to him, and to all those whose encouragement has stimulated me both to write and deliver these papers and to now present them in a new form.

TOM WRIGHT
St. Mary's College
St. Andrews, Scotland
New Year, 2014

SURPRISED *by* SCRIPTURE

I

Healing the Divide Between Science and Religion

WHEN I WORKED at Westminster Abbey, one of the questions most frequently asked by visitors, especially Americans, was "Is it true that Charles Darwin is buried here?" On one occasion, noting the route the visitor in question had just taken to walk through the abbey after evensong, I replied, "Madam, I think you just stepped on him." "Good!" came the emphatic reply, which told me something about the visitor in question. However, on another occasion, walking past Darwin's tomb, I spotted a little pile of flowers and greeting cards. They were obviously from schoolchildren, and the general tenor of their message was "Mr. Darwin, we love you."

I have often wondered what they had been taught. Can it really be that teachers tell the story of Western culture in terms of pre-Darwinian gloom, superstition, prejudice, and the dead hand of religion, with Darwin personally ushering in a new era of happiness, liberation, knowledge, and the milk of human kindness? If that isn't a highly selective and oversimplified history, I don't know what is. What's more, Westminster Abbey collects thousands of visitors from every corner of the world; how come it seems to be mostly Americans who are interested in him and who instantly take sides in an assumed war in which his very name is a battle cry?

I offer this short disclaimer. I realize that I am British (rather than American) and a theologian (not a scientist), and that I am therefore an outsider to this discussion. But I hope that I might point out three things in particular that an outsider may perhaps see more clearly than an insider.

First, I want to point out that the way the science and religion debate is conducted and perceived in North America is significantly different from the ways analogous debates are conducted and perceived elsewhere. Second, I want to suggest that this is at least partly because of the essentially and explicitly Epicurean underpinnings of the social self-understanding of the United States since the late eighteenth century—and that the standoff between science and religion in America is therefore analogous to, and indeed bound up at quite a deep level with, the standoff between church and state, or religion and politics, or however you like to put it, so that you can't address one of these topics without implicitly addressing all of them. Third, I want to propose that we therefore need a much more radical rethink of the underlying worldviews we are dealing with than we have normally contemplated in our science and religion discussions. That, I hope, is the point at which the deeper contribution of a biblical theologian might be useful.

The Raging Debate in North America

We in the United Kingdom never had a Scopes trial. We did, granted, have the notorious public debate at Oxford in June 1860, between Samuel Wilberforce, then bishop of Oxford, and the scientist T. H. Huxley. Within a generation, the story of this debate had grown and been shaped by a tradition so strong that that tradition has come to be accepted as true, though more recent research indicates that matters were by no means as clear cut as the received narrative would suggest.

According to the tradition, Wilberforce at one point asked Huxley whether he claimed descent from the apes on his grandfather's or grandmother's side, and Huxley retorted, more or less, that he would rather be descended from an ape than from someone who so abused his intellectual gifts. This, however, is the stuff of legend. The philosopher John Lucas pointed out some while ago, and this has been taken up by Stephen Jay Gould, that the account in which the agnostic Huxley struck the great blow for freedom from ecclesial dogma took root and spread at a time when the English middle classes, still anxious to gain political standing to rival the aristocracy, had a particular interest in the view that one's pedigree was irrelevant to one's moral worth.[1]

In addition, by the end of the century the world of science had changed significantly, from a sphere in which anyone and everyone might take part (provided they could afford it) to a much more professionalized guild. The picture of Free Science triumphing over the stuck-in-the-mud church fitted the increasingly independent mood of scientists in the 1890s, when some of the key texts about the Wilberforce/Huxley incident were penned, much better than the more freewheeling 1860s. But there, in the late Victorian era, the matter rested; and most people in today's Britain simply have a vague idea that the church is inclined to obscurantism, and that science has set us free from the shackles of its dogma and ethics. The horror of two world wars, with the Great Depression in between, gave people much more to worry about, and indeed far stronger reasons to question the roots of their traditional faith. So, to repeat my earlier point, my sense

---

1. J. R. Lucas, "Wilberforce and Huxley: A Legendary Encounter," available at http://users.ox.ac.uk/~jrlucas/legend.html.

today is that few people in Britain abandon their faith because of what science may say, though some who left it for other reasons or never had it in the first place naturally find it convenient to retell the stories not only of Wilberforce and Huxley but, farther back, of Copernicus and Galileo and the rest.

In America, however, the Scopes trial clearly had a massive impact, which resonated much more widely into the culture and accelerated a polarization that has not affected the rest of the world in the same way. I recently reread a devotional classic that I had much enjoyed in my early teens: Isobel Kuhn, *By Searching*. When Kuhn was a student in the 1930s coming from a Christian home, her science professors were scathing about anyone who believed not only in a literal six-day creation, but in any of the early stories in the Bible, and then more or less any of the Bible, including Jesus and Christian origins. The pressure from professors and peers to capitulate to the pushover science-therefore-atheism position was intense. And I suspect that, as usual, there was far more going on than simply a straightforward, rational—or even rationalist—public discussion. The modernist movement was in full swing culturally and politically, and it was assumed that Christianity, not only but not least its creation account and its belief in miracles, especially those of Jesus, was part of the premodern world that the forward move of history was leaving behind. In particular—and this, I stress, is an outsider's perception—it seems to me that the cultural polarization in American society, including the fundamentalist-modernist controversy of the first half of the twentieth century, has roots that go back at least as far as the Civil War in the 1860s.

Only that kind of backstory can explain the enormous interest generated across the United States by the Scopes trial of 1925, with a three-time presidential candidate (William Jennings Bryan) speaking for the prosecution and reporters flocking to Dayton, Tennessee, to cover the show. It was the first trial in the United States to be broadcast on national radio. But the Scopes trial cannot have generated the standoff between science and religion in and of itself; it merely brought it to a sharp, polarizing moment. And the trouble with sharp, polarizing moments is that they become iconic. Like martyrdoms ancient and modern, and indeed like civil wars, they generate loyalties and counterloyalties: you *must* now take such-and-such a line, because otherwise you're a traitor.

The words of Clarence Darrow, leading the team in Scopes's defense, have a contemporary ring for anyone who reads Richard Dawkins and his ilk: "We have the purpose of preventing bigots and ignoramuses from controlling the education of the United States." The vitriol of a leading journalist such as H. L. Mencken can hardly be accounted for except on the assumption that the trial was taking place at a major cultural fault line, with supposedly sophisticated city types from the Eastern Seaboard looking with disgust and horror at the "morons," "peasants," "hillbillies," and "yokels" of rural Tennessee. More significantly, subsequent accounts of the trial linked the antievolution mind-set to the rise of the Ku Klux Klan in the South. Whether or not that is sustainable as history is not the point. The point is that a great and painful wound in American society, which in its essence was not about science and religion but rather about the governance of the United States, the legitimacy of slavery, and the social place of African Americans, continued to fester and was given new depth and pain by the science/religion, or if you like evolution/Bible, debate.

This is why I say that, though of course the issues have been important elsewhere in the world, Americans seem to have had a particularly hard time of it, and that state of affairs continues to this day. After all, as is often pointed out, many leading conservative Christian theologians, in America as elsewhere, saw nothing particularly threatening in the discoveries of the nineteenth century. As with many other things, a path of wisdom that might have been available at an earlier stage was not taken, and instead the war between the two cultures—as in certain literal wars— continued, with attitudes on both sides hardening.

My point in this first section is that the present American context, which reflects these culture wars in newer forms, makes these issues much harder for Americans to deal with than they are for the rest of us. I am not saying there are no problems elsewhere; clearly there are. But when in the United Kingdom scientist-theologians like Alister McGrath or John Polkinghorne so clearly model ways of thinking in which the two worlds are wisely and richly integrated, most of us nonscientists are quite happy to continue that line of thought and see no need to trumpet our allegiances or to explain our conversions to new ways of thinking. These are not major cultural issues for us. They do not carry—as I fear they often do in America—worryingly direct political implications. Clearly the American issues are important. But it may help to reflect on how they are bundled up with larger issues, gaining a lot of their apparent heat from those larger problems rather than from their own innate difficulties. And I say all this not simply in the sense of "we don't see it quite like that." I say it because there is a danger of Americans assuming that everyone else *ought to* share their problems. There is an old story about a gang of youths in Belfast stopping an Indian gentleman on the street. "Are you a Catholic or a Protestant?" they demand. "I'm a Hindu!" he answers. "Okay," they say, "but are you a Catholic Hindu or a Protestant Hindu?" So, am I a fundamentalist creationist or an atheistic scientist? Answer: I'm a Brit.

The Problem of Epicureanism

There is a much longer backstory that has affected, and afflicted, both Europe and America over the last two or three centuries. I refer to the rise and enormous power of the philosophy that was originally called Epicureanism. This, basically, is the worldview in which God, or the gods, may perhaps exist, but if they do, they are far away and remain uninvolved with the world.

(The difference between Epicureanism and Deism, broadly, is that a Deist god probably made the world in the first place but then retreated, whereas an Epicurean god certainly didn't make the world and has not been involved with it since.) God lives at the top of the building, and we live at the bottom; the stairs have been destroyed, and the elevators stopped working a long time ago.

The essential point is this: the whole project we know as Modernity, with the European and American Enlightenment movements as its flagships, was based on Epicureanism, and the effects of this are apparent not only in questions of science and religion but in many other areas of life, not least the political. And here is the shock for many today: what is commonly thought of as the post-Darwin world of physical science is not a new discovery. The celebrated claim of the Enlightenment, to have inaugurated a new *saeculum* (which is after all what it says on the dollar bill: *"Novus ordo seclorum"* is one of the most breathtaking claims ever made by a young country, parallel I suppose to the French Revolutionaries' restarting of the calendar; their experiment didn't last, whereas the dollar bill still proclaims the new age)—this celebrated claim of the Enlightenment, to have launched an entirely new worldview on the basis of fresh scientific discoveries, is simply bogus. What happened was that the push to Epicurean philosophy on many fronts had been gathering energy for quite some time, and the new discoveries were hailed as evidence for it.

Epicureanism Defined

What exactly is Epicureanism? Epicurus himself was a third-century BC Greek philosopher. He was fed up with the murky world of pagan religion in general and the Stoic pantheists in particular. He argued that the gods did not concern themselves with our world, whether to intervene in it or to judge its inhabitants after death. In fact, since the physical world continued on its self-caused way without help from outside, physical death simply meant complete dissolution of the human being. You could sum up Epicurus's philosophy, at least in its desired effects, with the slogan that Richard Dawkins and his associates put as advertisements on London buses two or three years ago: "There's probably no god. Now stop worrying and enjoy your life." The paganism of ancient Greece told tales, remarkably similar to some would-be Christian tales, about the gods being angry with you in the present life and threatening to burn you alive after death. No, declared Epicurus, the natural world goes on its way by the random collision of atoms. (He didn't mean by "atoms" exactly what we now mean, but that doesn't matter much for our purposes.)

So, for Epicurus, there was nothing to worry about. Draw a direct line from him to John Lennon: imagine there's no heaven, no hell beneath us; now get on and live for today. The image of Epicurus as a hedonist is true, but it was a very refined hedonism, since he taught that the more obvious bodily pleasures didn't last and often produced less pleasurable side effects. (As an aside on the bus advertisement: In October 2011, the American apologist William Lane Craig had arranged to debate Richard Dawkins at the Sheldonian Theatre in Oxford, but Dawkins withdrew from the debate late in the day. The organizers then put posters on the Oxford buses saying, "There's probably no Dawkins. Now stop worrying and enjoy October 25th at the Sheldonian Theatre.")

The philosophy of Epicurus was given a major new lease on life by the Roman poet Lucretius, who lived about seventy years before Jesus. Like his master, Lucretius was fed up with the popular Roman religion, which induced fear of the gods, and particularly fear of death and what might happen afterward. In an astonishing poetic masterpiece, *De Rerum Natura Concerning the Nature of Things,* Lucretius expounded Epicurus's ideas, adding some of his own but basically taking the project forward in a manner calculated to appeal to the intelligent and aesthetically inclined Roman world of his day. (Was that the last time a major philosophy had appeared in the form of a long poem? I don't know. Would Dante count?)

In Lucretius it all becomes clear and straightforward. The world is what it is because of (what he called) atoms, which, free-falling through space, collide with one another, sometimes combining and sometimes bouncing off. There are clearly different kinds of atoms, which is why they have these different effects; major changes are caused by the inexplicable "swerve" that sometimes happens to the atoms so that they veer off in new directions and produce different results. But the main point is essentially what we would today call the evolutionary thesis: life in the world has developed under its own steam as the random by-product of chance collisions and combinations of atoms and the more com-plex life-forms they produce. The gods are out of the picture. They are nowhere to be seen. Death is, quite literally, nothing at all: the atoms disperse, never to recombine.

As a historical side note, I should say that in the ancient world it was much easier to be an Epicurean if you were fairly well-off. If you were an ordinary lower-class person—that is, among the 95 percent of the free population—or a slave, the exhortation to relax and enjoy your life might have rung somewhat hollow. It is interesting that Christianity thrived precisely among those for whom the major existing philosophies had little to offer.

Perhaps that is why, with the new Europe emerging from the Middle Ages in the fifteenth and sixteenth centuries, and boast-ing a burgeoning middle class with enough prosperity to educate itself, Epicureanism emerged from the shadows to which it had been consigned by the Christians and Jews of late antiquity and made a remarkable comeback. The story has recently been told by Stephen Greenblatt in *The Swerve*. The subtitle, *How the Renais-sance Began,* is an overstatement: the Renaissance had many roots. However, Greenblatt makes a good case that the rediscovery of Lucretius in 1417 was a seminal moment. In the Europe of the late Middle Ages there were many people who, like Epicurus and

Lucretius, were fed up with the frightening and bullying theology they had been taught. Supposing that if there was a god, he was detached from the world, that the world was not invaded and manipulated by angels and demons, and that human life was something to celebrate in its own terms, was a very attractive alternative. Curiously, the man who found the manuscript, Poggio Bracciolini, did so in 1417, exactly a century before Martin Luther tumbled upon a version of Paul's doctrine of justification by faith, enabling him, too, to similarly reject the idea of an angry, vengeful God. (Or did he?)

That imperative—to get away from the bullying boss in the sky and do things differently as a result—continued to inspire a long line of Epicurean thinkers, from Machiavelli to Hobbes, through to the great thinkers of the Enlightenment and ultimately to Thomas Jefferson himself. Indeed, the Enlightenment was, as a whole, one long determination to get rid of the big, bad boss upstairs. That is why one of its main drivers was the Lisbon earthquake of 1755. Had there been a god who was running the show, he certainly wouldn't have allowed such a thing, on All Saints' Day in particular, when everyone was inside the collapsing churches. So, with Voltaire and others, Europe pushed God upstairs out of sight, and many in America followed suit.

This is a fascinating story, though much longer and more complicated than there is time or need to tell here. I want to focus on two things in particular. First, for Epicurus and Lucretius, the main point was that the theory about atoms colliding and swerving was central, because it left no gap in the creation and causation of the world through which divine beings might intrude. The world made itself without intervention. This kind of atomism was already making its way into the bloodstream of European culture not only well before Charles Darwin set sail on the *Beagle* but well before his grandfather Erasmus articulated early versions of what became evolutionary theory, in the late eighteenth century from his home in Lichfield, almost next door to the deanery where I lived for five years.

Not only was the theory of evolution not invented by Darwin himself; it wasn't invented by his Enlightenment predecessors either. It was invented by Epicurus (looking back to Democritus and others) and popularized by Lucretius. It wasn't and isn't a new, modern discovery. It is simply one part of one ancient worldview. It has enjoyed waves of popularity, the current one so widespread and long lasting that it's easy for many, not least many of today's scientists, to suppose it is the only possible one. It isn't; I will say more about that presently. But the verdict of the philosopher Catherine Wilson is spot on: in the modern world, "we are all Epicureans now." It is the default mode, sadly, for most Christians who oppose modern science as well as for scientists who oppose modern Christianity. That is the problem behind all the specific hand-to-hand fighting over particular issues. There is still enough residual Judaism and Christianity in the culture for people to remember rumors of a different view, and there are hints too at a Stoic pantheism in which God is omnipresent. But for most people the Epicurean view, that God is a long way away and stays out of touch, is the reality. Emily Dickinson summed up this double vision in a famous line (in a letter to Mrs. J. G. Holland [L551] in spring 1878): "They say that God is everywhere, and yet we always think of Him as somewhat of a recluse." In other words, the rumor of Stoicism is not dead, but the reality of Epicureanism is how we live.

The second point I want to make about the rise of Epicureanism at the dawn of modernity, and particularly in the origins of the Enlightenment, is that it was seized upon not least because of its *political* implications. This is clear already in Machiavelli and Hobbes, but it comes to the fore in the late eighteenth century. In parallel with the idea that atoms simply do their own thing, falling through space, swerving this way and that, colliding and combining with one another but importantly *without outside control,* specifically without divine control or intervention, the Enlightenment thinkers were busy rejecting the political ideologies they had inherited, in which nobles, and particularly monarchs, basically controlled things, and the lower orders just had to put up with it.

The rejection of the big bully in the sky and reliance on an atomism in which the atoms do their own thing without influence from above have an obvious analogue in the rise of what we now think of as liberal democracy. Gone is the divine right of kings, swept away with the guillotine in France and the Boston Tea Party in America. *Vox populi vox Dei* is the cry—but then *Deus* himself disappears off into the far beyond, and *vox populi* is all we're left with. Welcome to the brave new world of the nineteenth century. No wonder evolutionary biology got such a fair wind. It pressed all the right buttons. It proposed that the natural world was not static but in a state of flux, which was how many people wanted to see the social world as well. *The Origin of Species,* after all, appeared in 1859, only eleven years after the "year of revolutions" in 1848. Nobody listened to the protest of the eighth-century Alcuin, that *vox populi vox Dei* didn't work because a riotous crowd is a form of madness. The madness that bothered the Boston rioters was that of King George III.

Of course, the weaknesses of Epicureanism as a philosophy show up in the political realm, perhaps more obviously than elsewhere. Alcuin was right. The political equivalents of falling and colliding atoms have not always been a pretty sight. As a (small-*d*) democrat myself, I would quickly say that the tyrannies which preceded modern democracy and the tyrannies that defaced the twentieth century were not pretty sights either. Democracy can generate new forms of tyranny, and once we have sold our souls to a particular voting process there is no way back. We need to return to the drawing board and think more clearly about whether the natural and proper human passion for freedom and the natural and proper human need for order and stability are best served by the kinds of democracy we have developed, without the aid of divine or monarchical intervention from above, on the model of the Epicureanism that has proved so popular and influential.

The Relationship Between Epicureanism and Science

I hope it is obvious where all this is going in relation to the current American situation. First, some reflections on the relationship between ancient Epicureanism and modernist science. It may well be the case, just as fifteenth-century Epicureanism opened the way for aspects of the Renaissance, that Enlightenment Epicureanism opened the way for questions to be asked from new angles. But whereas it has been assumed, for instance, that the discoveries of Charles Darwin and others proved the worldview with which they had started—which was not, I stress, a modern worldview, but only a modern version of an ancient one—the conclusion is not warranted by the evidence. Let's leave aside the problems that some still suggest are latent within a Darwinian account of the way the world is. Supposing it all works, it does not follow that the Epicurean worldview, with absent gods and independent atoms, is correct. That would be the case only if causation were a zero-sum game, so that *either* God *or* observable physical causes were involved. As soon as you challenge that rather naïve assumption, all sorts of other options are open.

This is the point, in fact, at which some of the sharpest and angriest questions are still asked. If you are supporting Darwin, a furious correspondent wrote me the other day, that means you don't believe in miracles—so you can't really believe in the resurrection or the virgin birth or whatever else. Now a central part of my problem with this whole discourse is that the very word *miracle* itself, in the way we now hear it in post-Enlightenment Europe and America, is bound to be fatally damaged by the implicit Epicureanism of our latent worldview. So too with the word *supernatural,* which was used well before the Enlightenment but since then has taken on resonances of the same worldview. The problem could be put like this: in science/religion debates, or evolution/creation debates, it is all too easy for the scientists or evolutionists to state their position in Epicurean terms, *and for the Christians or creationists to follow suit.* Of course, the Christians or creationists wouldn't actually be Epicureans, precisely since they believe in a god who did make the world and who does still run it. But they inherit and operate within the deeply damaged vision of the creator and the cosmos that they get from Deism, and which shares its worst features with Epicureanism: that some things happen naturally, while other things happen only because God makes them happen.

A striking example of this, which I wrote up in my book on virtue (*After You Believe*), occurred in New York in January of 2009. An aircraft took off from LaGuardia and almost at once ran into a flock of geese. The pilot, Chesley Sullenberger III, made several lightning decisions and performed dozens of complex flying maneuvers in a couple of minutes, and the plane landed safely on the Hudson River. Lots of people said it was a miracle, and I wouldn't for a moment say that God was not involved in that whole process. But the reason the plane landed safely was that Sullenberger had been flying planes and gliders, and teaching others to do so too, for thirty years. His character had been formed so that all those complex thoughts and actions were second nature. The danger in using the word *miracle,* in other words, is that we assume the zero-sum either/or. *Either* God did it *or* the pilot did it. And it is that assumption, shared by post-Enlightenment Christians and secularists alike, which needs to be challenged in the name of a genuinely biblical worldview.

Second, some reflections on the political context. One of the biggest problems I think Americans face is that the science/religion debates take place in a world where the analogous political situation is entrenched in everybody's minds and hearts. America has instantiated in its political system the most explicit form of Epicureanism ever, I think, to reach the written constitution of a country. God and the world don't mix; faith and the public square, religion and education, prayer and schools—you name it. And the great democracy goes on its way, like Lucretius's atoms, under its own steam, with occasional inexplicable swerves and plenty of collisions, combinations, and even combustion of the political equivalent of those atoms. Of course, given the way things actually work, it would be hard to maintain the fiction of a causeless democracy with individual citizens making up their mind on real issues. Were that so, I suspect that presidential hopefuls would stop spending millions on their campaigns.

Now, of course, there are many who have tried to move back toward a fresh integration. This has sometimes been done for merely pragmatic reasons, seeing that in fact you can't keep God out of the public square, or public issues out of church life. Sometimes it has been done for theological reasons, seeing that the God of the Bible does in fact claim sovereignty over every aspect of human life. It used to be the case that the Christian Right in America was criticized for trying to bring religion back into politics, while in the United Kingdom it was mostly the Christian Left that was doing it. Things are a bit more complicated now on both sides of the Atlantic. But American culture, far more than British culture, militates against the mix.

My point is this: if you're trying to have a discussion about God's involvement in the world in one area—creation, science, whatever—while living and breathing a system in which God has been disinvolved with the world by definition and by act of Congress, there is an opposition set up, deep within the structure of how people think, that is going to make it very difficult. That is why, I think, some of those who insist on God's actions in creation and providence, who see him as a God who is essentially outside the whole process and who reaches in, despite the Epicurean prohibition, and does things for which there was otherwise no cause, sound quite shrill. They are desperately insisting on the truth of something that, at a structural and presuppositional level, has been ruled out of court, declared unconstitutional.

A further social and political oddity emerges at this point. It is well known, and the opponents of Darwinism regularly point this out, that social Darwinism has been very powerful in Western culture. Actually, something like social Darwinism was alive and well long before Darwin himself, and I have read differing reports on whether Darwin himself either was attracted to it or actually embraced it. But there is no doubt that social Darwinism was rampant in Europe and perhaps also in America in the early twentieth century. That was one reason why most people in Europe seemed to think it was good to have a big war now and then, to purify the race and allow the fittest to survive. Of course, the First World War showed the utter stupidity of that: machine guns kill the most highly sophisticated, intelligent, and strong young men at random. But that hasn't stopped an implicit social Darwinism from entrenching itself deeply into American culture, perhaps in particular out of all the post-Enlightenment cultures.

Basically, the American dream is that if you get up and go, you'll succeed; the egalitarian hope is that the fittest will survive the economic jungle. This is simply a given, an unexamined presupposition that lies behind, for instance, the gut-level reaction against any kind of health-care proposal: after all, if these folks were fit to survive, they'd be out there earning a living! It also works at the international level: America has grown to be the leading superpower, so if America doesn't like a regime somewhere else in the world, then America—with a tiny bit of help from her friends, of course!—has the right and duty to go and bomb it and effect regime change. And my point, as you will readily see, is the great irony that often those who are most opposed to Darwin when it comes to reading Genesis 1 are in fact most deeply in thrall to him, or to the wider application of his theories, when it comes to social and international policy.

It is thus clear to me that the entire modern Western world has been deeply soaked in the philosophy of Epicurus and Lucretius. This has been all the more important in that it has gone unrecognized, and people have been taught instead from the cradle—in some cases, from the cradles of emerging nations!—that the new science and technology of the eighteenth century ushered in a new world whose participants have been elevated to a new level of human living, a new *saeculum* whose inhabitants have grown up, come of age, and can now look back on their former ignorance with a mixture of pity and horror. Within this Enlightenment worldview—which of course gains its plausibility from the undoubted fact that we *have* indeed discovered millions of things we never knew before, from DNA to the atom bomb, from air travel to microchips—people who know nothing of ancient philosophy are taught, *as though it were a new thing,* that gods may exist but if so they remain a long way off, or perhaps that the rumor of their existence was greatly exaggerated in the first place. And many who *do* know ancient philosophy are quite happy that Epicureanism should come into its own at last as the philosophy no longer of a small elite but of the masses.

Sometimes, indeed, one hears would-be Christian thinkers hailing the new world as a great liberation and a new opportunity for the faith. A recent article by a leading Jesuit thinker celebrated evolution without any attempt to come to terms with a Christian vision of it, but rather using it as a way of getting rid of several things in traditional Catholic culture of which he clearly disapproved, from Augustine's view of original sin to Anselm's view of atonement in terms of satisfaction, with consequences all down the line from sacramental theology to women priests and sexual ethics.[2] It is this kind of thing, of course, that produces an equal

---

2. Jack Mahoney SJ, "Humanity's Destiny," *The Tablet,* January 14, 2012, 7–8.

and opposite reaction, with people in many Christian traditions saying if that's what accepting evolution does for you, then we obviously have to resist it for all we're worth.

But my case is that once we recognize the deep-rooted Epicureanism of much of modern Western culture, our vocation as Christian thinkers is not to make an easy, compromising peace with it but to discern how to restate and reinhabit a genuinely Christian worldview in its place. The real problem ought not to be posed in terms of faith on the one hand and science on the other, but in terms of a worldview that splits faith and science (held, and trumpeted shrilly, by creationists and Dawkinsians alike) and one that does not. This ought really to be the subject of a whole book or series of books, but in the remainder of this chapter I simply want to sketch some features of such an attempt.

A Christian Response to Epicureanism

So what are we to say to all this? The good news is that the philosophy which has prevailed in the last two hundred years was well known in the ancient world as well, and writings of the church give us the clues by which we can develop an appropriate response to it.

The Early Christian Response

The earliest Christians, being mostly poor and unsophisticated, did not develop a detailed and thought-out response to any of the prevailing philosophies of the day. We catch hints of such debates in Paul's letters and in his famous address on the Areopagus, but it is left to the apologists of the second and subsequent centuries to begin the high-octane engagement with the philosophies of Greece and Rome. In that process, as Jewish thinkers also found,

Epicureanism was simply ruled out. To this day the Hebrew or Yiddish word *apikoros* is used to mean "a heretic," someone who has denied the basis of the faith. Because, of course, the fundamental doctrine of Judaism and Christianity was then and is still the belief that the world we know is the good and wise creation of a good and wise God. The Jewish people hail this God as the God of Abraham, Isaac, and Jacob. The Christians, sharing this, go farther: the creator is the father of Jesus the Messiah, the Lord.

The early Christians, even Paul, did not develop a detailed doctrine of creation. They did not need to, since they inherited one: the ancient biblical vision of Genesis. But, as they well knew, the Genesis account is a highly poetic, highly complex narrative whose main thrust has nothing to do with the number of twenty-four-hour periods in which the world was made, and everything to do with the wisdom, goodness, and power of the God who made it. And one feature of the Genesis account that always strikes a nonspecialist like me is the fact that God is said to have made a world *that will then make itself*: trees and plants that bear fruit containing seed for more of the same, animals that reproduce after their kind, and above all humans, called to reflect God into the world and the world back to God, who likewise are to be fruitful and multiply, to be dependent cocreators.

And just as the account of the Fall in Genesis 3 sits alongside the account of the rebel angels in Genesis 6, teasing the wise reader into pondering evil as a genuine mystery rather than an easily explicable (and soluble!) glitch, so the accounts of creation itself in Genesis 1 and 2 do not sit neatly on top of one another but offer very different angles of vision. The fact that the animals are created before the humans in Genesis 1 and the male human before the animals in Genesis 2 is a classic literary way, perhaps a classic Hebrew literary way, of saying that these two accounts are signposts pointing away from themselves to a third reality that remains unstated, perhaps unstatable. Perhaps it is that farther reality to which the Psalms and Proverbs are pointing when they speak of God making the world "by wisdom." And perhaps it is that farther reality that Paul, John, and Hebrews refer to when they pick up the hint and speak of Jesus himself in the language of wisdom, as the one through whom all things were made.

Genesis and the Christian Worldview

Genesis 1 as it stands would have been seen in the ancient world as a story about a god building a *temple,* a place for his own habitation, into which he would of course place an image of himself before coming to dwell in it, to take his ease there, to be at rest.[3] The temple in question is the combined heaven-and-earth reality, a construction in which heaven and earth are not, as in Epicureanism, separated by a great gulf, nor, as in Stoicism, fused together so that they are in effect the same thing. God is present in the garden, close, intimate, walking there at the time of the evening breeze in Genesis 3. It is a world where God remains God and humans remain creatures, but in which God's sphere, heaven, and the human sphere, earth, interpenetrate and are, by design and in fact, mutually permeable.

One explanation for the phenomena that give rise to Epicureanism is that after Genesis 3, however we interpret it, the sense of alienation leaves one with an either/or: either admit that something has gone wrong and seek by whatever means to restore the relationship, or develop a theory that says there always was a great gap between the gods and the world, postulating an ontological gap where Genesis sees a moral and spiritual one. But the point is not to explain it but to answer it. Stoicism answers Epicureanism by denying the problem: divinity is present in the world and in ourselves, and if we don't like it that's our problem. But Judaism

---

3. John H. Walton, *The Lost World of Genesis One: Ancient Cosmology and the Origins Debate* (Downers Grove, IL: InterVarsity, 2009).

and Christianity classically answer both philosophies by continuing to celebrate and explore the mysterious interpenetration of heaven and earth. One of my favorite verses in the Psalms is from Psalm 65, celebrating that moment of exultation when the rich beauty of early morning and late evening are seen as actual celebrations of the creator's wisdom:

> *By your strength you established the mountains;*
> *you are girded with might.*
> *You silence the roaring of the seas,*
> *the roaring of their waves,*
> *the tumult of the peoples.*
> *Those who live at earth's farthest bounds are awed by your signs;*
> *you make the gateways of the morning and the evening shout for*
> *   joy.* (Psalm 65:6–8, NRSV)

The psalm goes on to speak, in tones reminiscent of Genesis 2, of God's river being full of water, producing the bounty of harvest and the celebration of pastures, hills, meadows, and valleys. The older I get, the more I realize that this is not just a pretty way of speaking. Jew and Christian alike are charged with reminding one another of the ceaseless hymn of praise that ascends to the creator from the inanimate creation, as the heavens declare his glory and the firmament proclaims his handiwork. He feeds the young ravens when they call upon him; contemporary science might attribute the raven's cry for food to what we call instinct, and I wouldn't disagree, but talking about "instinct" within an Epicurean framework always implies reductionism. "The world is charged," said Gerard Manley Hopkins, "with the grandeur of God," and Hopkins clearly was no pantheist.

Rather, what we find throughout Old and New Testaments is a rich theology of creation in which the whole earth was *already* full of the divine glory—that, after all, is what the seraphim were singing in Isaiah 6—and also in which the whole earth *would in the future* be filled with the knowledge of YHWH, or with the knowledge of the glory of YHWH, as the waters cover the sea (Isaiah 11:9; Habakkuk 2:14). This notion of filling the whole world with the creator's glory joins up with the vision of creation as a temple. The first great cycle of narrative in the Bible is that which runs from Genesis 1 and 2 to the end of Exodus, where, despite Israel's idolatry and sin, the creator and covenant God nevertheless comes to dwell in the newly constructed tabernacle, filling it with his glorious presence (Exodus 40). When, long afterward, the Jerusalem Temple was built, the same thing happened (1 Kings 8). That presence departed at the time of the exile, but the prophets assured Israel that the divine splendor would return one day; the New Testament writers declare, in many different ways, that it has happened—in Jesus the Messiah and through the gift of God's spirit. Jesus himself is the true temple; then, by extension as it were, those who are gifted with his spirit become, in themselves, living temples.

This is how Isaiah's prophecy is fulfilled, that the glory of the Lord shall be revealed for all flesh to behold—the promise of a new temple, a new creation, in which all will be invited to share. It has all come true, says the New Testament, in Jesus, and then in the Spirit. And this means that everything we might say about the relationship between heaven and earth, between God and the world, between faith and science, between piety and public life—all those analogous questions that have so baffled Western modernism—find their answer in Jesus. But saying that by itself doesn't (in my experience) help much. It sounds pious, and hard for a Christian to quarrel with, but remains impossibly and incomprehensibly dense. What does it actually mean? Others have suggested Christology as a way of reimagining how the worlds of science and faith can be drawn together. I make the same suggestion but within what seems to me a more fully biblical approach to Christology itself.

Jesus Ushers in a New Creation

This means, to begin with at least, that we have to read the New Testament's big claims about Jesus and creation the other way around. We have read John 1, 1 Corinthians 8:6, Colossians 1, and Hebrews 1 as though they are simply making grand claims about Jesus; as though we know what creation is, and we now discover just how exalted Jesus is by being told that "all things were made through him." But, as with Christology in general, so with the Christological claims about creation: perhaps we don't after all know what creation is, just as we don't after all know in advance who God is; perhaps what we are being told in these famous passages is that we are to search for an alternative account of creation, a cosmogony, by more fully pondering Jesus himself.

There we run into another huge problem, since Western writing about Jesus has been drastically distorted precisely by the Epicureanism that we are seeking to controvert. Jesus has been pulled and tugged this way and that, by naturalists who want to make him into another great religious teacher (or perhaps another great political revolutionary) and supernaturalists who want to make him the divine "invader" from "beyond," performing miracles to prove his supernatural power and summoning us to leave this world and return with him to his. These pictures simply reflect the false either/or of Epicureanism, as does, particularly, the pseudoeschatology of the rapture in dispensationalism. None of this begins to do justice, historically or theologically, to what the four Gospels are saying.

The four Gospels, completely in line with Genesis, the Psalms, Isaiah, and the rest, tell of *how God became king:* how the creator God, in and through Jesus of Nazareth, launched his new-creation project for the world. On every page, what we are looking at is precisely new *creation:* not just a new spirituality, certainly not a system for rescuing people *from* this world, but a movement of God's creative spirit, anointing Jesus but also breathed out by him, through which humans are called to become genuine humans at last, rescued from all that thwarts that, and so equipped to carry forward God's plan of new creation. Once again, God is remaking the world not by an intervention that drowns out everything else, as some supernatural schemes would have it, nor by allowing natural causes to take their course, as some evolutionary schemes—including some would-be Christian evolutionary schemes!—would have it, but by the act of redemptive new creation through which humans are able once more to reflect God into his world and the world back, in worship, to God. The whole project of Jesus is a *new-temple* project, which is why the Jerusalem Temple and then the pagan temples become so problematic in the Gospels and Acts; it is the project, in other words, in which *heaven and earth are brought together at last,* with God's sovereign rule extending *on earth as in heaven* through the mission of Jesus, climactically in his death and resurrection, and then through the similarly shaped and spirit-driven mission of his followers.

My proposal, then, is that if we want to make the real quantum leap that the science-and-religion debate badly needs, we should look deeply into the four Gospels and their story of Jesus inaugurating God's kingdom, dying on the cross, and rising as the firstfruits of the new creation, and ask ourselves about the nature of the new temple, the new heaven-and-earth reality, and the new creation itself, which Jesus was modeling and launching. Within that reality, it wouldn't simply be a matter of finding a way to reconcile atheistic science with rationalistic Christianity, either by letting them live in separate spheres or finding some kind of awkward in-between accommodation. That simply perpetuates the split-level Epicurean worldview that I have argued has been the problem all along, just like the studies of Jesus that make him either Superman or "just another great teacher."

But if we're talking about reimagining and relaunching the Christian worldview in the Western world, it isn't only science and religion that have to be thought through. It is the whole of the way we do society and politics, personal life, and not least, mutual responsibility across the global family. If we pay attention to Jesus as the pattern and beginning of new creation and thus learn what it means that "in him all things were created"—in other words, if we search seriously and genuinely for a Christian cosmogony—then we can expect to be challenged on other levels as well. But that, after all, is what we, as followers of Jesus—and as followers of true science—ought to expect.

2

Do We Need a
Historical Adam?

T HE ROOT PROBLEM we face as Christians is that in articulating a Christian vision of the cosmos the way we want to do, we find ourselves hamstrung because it is assumed that to be Christian is to be anti-intellectual, antiscience, obscurantist, and so forth. This constitutes a wake-up call to us in this form: though the Western tradition and particularly the Protestant and evangelical traditions have claimed to be based on the Bible and rooted in scripture, they have by and large developed long-lasting and subtle strategies for not listening to what the Bible is in fact saying. We must stop giving nineteenth-century answers to sixteenth-century questions and try to give twenty-first-century answers to first-century questions. Our concern is for the truth and beyond that for our love of the God of truth and our strong, biblically rooted sense that this God calls us to celebrate the wonder of his creation and to work for his glory within it.

There are two theological drivers for people to believe in a young-earth creationism and a historical Adam. The first supposes that if people let go of this position, they are letting go of the authority of scripture. I suspect, myself, that sociocultural factors are among the main influences. In dispensationalism in particular, a flat, literal reading of Genesis is part of a package that includes the rapture, Armageddon, saving souls for a timeless eternity, and so on, together with the usual package of ultraconservative (as it seems to a Brit) policies in society, government, and foreign policy. So I suspect we need to think through the question of how the authority of scripture actually works and what it might mean in this case.

But there is a second theological driver of the problem. This has to do with the deep-rooted Western soteriology that has characterized Catholic as well as Protestant, liberal as well as conservative: a sense that we know, ahead of time, that the Bible, particularly its central New Testament texts like the Gospels and Romans, must really be about the question of how we get saved. For some, particularly in the Reformed tradition, the question of Adam as the federal head of his descendants is one part of the soteriology that sees Jesus the Messiah as the federal head of all those who are "in him." So let me say something brief about scriptural authority, and then something slightly fuller about Adam and related issues.

Biblical Authority

In my book *Scripture and the Authority of God,* I outline a fresh way of talking about biblical authority that is rooted in the Bible—in contrast, I have to say, to some dogmatic schemes that say a lot about authority or inerrancy or whatever but seem to pay remarkably little attention to what the Bible itself is actually about. I develop this view in conscious dialogue with the conservative position, which seems to me to go like this: it's either the Bible or the pope, so it must be the Bible, so we have to stand by every letter of scripture or Catholicism will swallow us up. That then turns round to face the rationalist or secularist challenge with the same position: the Bible must be literally true from top to bottom, or it all collapses into a mess of woolly liberalism with no gospel, no morality, and no hope. But simply saying, "The Bible is the only authority," is not enough. We have to nuance it, and when we do, an interestingly different picture emerges. Let me sketch it extremely briefly and refer you to the book for a fuller treatment.

In the Bible all authority belongs to God and is then delegated to Jesus. The risen Jesus doesn't say, "All authority in heaven and earth is given to . . . the books you chaps are going to go and write." He says, "All authority has been given to *me*." The phrase *authority of scripture* can only, at its best, be a shorthand for *the authority of God in Jesus, mediated through scripture*. Why would we even want to mention biblical authority? Why not say, "We live under Jesus's authority," and leave it at that? Wouldn't that be the biblical thing to do? Well, yes, but as centuries of history demonstrate, *the Bible is the God-given means through which we know who Jesus is*. Take the Bible away, diminish it or water it down, and you are free to invent a Jesus just a little bit different from the Jesus who is hidden in the Old Testament and revealed in the New. We live under scripture because that is the way we live under the authority of God that has been vested in Jesus the Messiah, the Lord.

But what is God's authority there for? Certainly not to give us a large amount of true but miscellaneous information. Solomon made lists of natural phenomena, but they didn't get into the Bible. The Bible is not an early version of the *Encyclopædia Britannica*. Here is the point that will be central to the second half of this chapter: the point about God's authority is that the whole Bible is about God establishing his kingdom on earth as in heaven, completing (in other words) the project begun but aborted in Genesis 1–3. This is the big story that we must learn how to tell. It isn't just about how to get saved, with some cosmology bolted onto the side. This is an organic story about God and the world. God's authority is exercised not to give his people lots of true information, not even true information about how they get saved (though that comes en route). God's authority, vested in Jesus the Messiah, is about God reclaiming his proper lordship over all creation. And the way God planned to rule over his creation from the start was *through obedient humanity.* The Bible's witness to Jesus declares that he, the obedient Man, has done this. But the Bible is then the God-given equipment through which *the followers of Jesus are themselves equipped to be obedient stewards, the royal priesthood, bringing that saving rule of God in Christ to the world.*

Therefore, the Bible does what God wants it to do when, through the power of the Spirit, it enables people to believe in Jesus, to follow him, and to share the work of the kingdom— not building the kingdom by our own efforts, of course, but, as I say in *Surprised by Hope,* building *for* the kingdom. We become sharers in God's kingdom work by loving him with heart, mind, soul, and strength, and the Bible is the primary means the Spirit uses to bring about that heart-and-life renewal. The authority of scripture is therefore the *dynamic,* not static, means by which God transforms humans into Jesus-followers and therefore kingdom-workers.

One of the wonderful things about the Bible is the way no generation can complete the task of studying and understanding it. We never get to a point where we can say, "Well, the theologians have sorted it all out, so we just put the results in our pockets or on the shelves, and the next generation won't have to worry—they can just pull it out and look it up." No, the Bible seems designed to challenge and provoke each generation to do its own fresh business, to struggle and wrestle with the text. I think that is the true meaning of the literal sense, in Augustine's sense of "what the writers really meant": we have to acquire those old eyes, the historian's quest to understand Genesis and Matthew and Romans in their historical context. I know that is strongly resisted today by many conservatives, but this is ridiculous: without historical inquiry, parallels, lexicography, and so on, we wouldn't even be able to translate the text. And, yes, I know that there are many secularizing biblical scholars, and indeed many left-brain-dominated conservative ones, who produce a kind of biblical scholarship that the church either shouldn't use or couldn't use. But just because the garden grows weeds, that doesn't mean we shouldn't plant fresh flowers, instead paving the whole thing over with concrete. No, each generation must do its own fresh historically grounded reading, because each generation needs to *grow* up, not simply to *look* up the right answers and remain in an infantile condition. This too is part of kingdom work.

Finally (and here I draw your attention to my book *How God Became King*), the problem with all hand-me-down solutions, and especially the rules of faith and even the great creeds, is that they have screened out the central Biblical message, which is the coming of God's kingdom. There is nothing wrong with the creeds and the rules themselves. What they say is true. But they oversimplify, and when people then start to build systems on that oversimplification they miss the central point. People today sometimes talk about canonical readings of scripture, meaning classic orthodox readings; but classic orthodoxy has routinely forgotten that the central message of the Gospels, as of Jesus himself, was that through him and his work and his death and resurrection, the living God was becoming king on earth as in heaven. If we aren't getting that message out of the Bible, we aren't reading the Bible itself but rather allowing our traditions to echo off the surface of a text that is trying to tell us something else. Rules of faith and creeds are like the guard rails on the side of the highway, which prevent you from skidding off into the path of oncoming traffic. They do not themselves tell you everything you need to know about your journey and destination, nor do they put fresh gas in your tank. Only the scriptural message about God's kingdom in Jesus Christ will do that.

All this, I think, is what it means for me to call myself an evangelical, and I grieve that that word has so often meant a closing down of scripture reading rather than its opening up. The Bible is there to give you the questions and the agenda, to shape you into being the people with courage and skill to answer those questions and follow that agenda. All too often the word *biblical* has been shrunk, so that it now means only "according to our tradition, which we assume to be biblical." And as Paul asks in 1 Corinthians 8 and Romans 14, how do you tell which things we can hold without dividing our fellowship?

I wonder whether we are right even to treat the young-earth position as a kind of allowable if regrettable alternative, something we know our cousins down the road get up to but which shouldn't stop us getting together at Thanksgiving. Yes, of course any confrontation must be done in courtesy and civility, charity and gentleness—though if the truth is at stake, look at how Paul confronted Peter in Antioch. . . . And if, as I suspect, many of us don't think of young-earthism as an allowable alternative, is this simply for the pragmatic reason that it makes it hard for us to be Christians because the wider world looks at those folks and thinks we must be like that too? Or is it—as I suggest it ought to be—because we have glimpsed a positive point that urgently needs to be made and that the young-earth literalism is simply screening out? That's the danger of false teaching: it isn't just that you're making a mess; you are using that mess to cover up something that ought to be brought urgently to light.

I think what has happened is this. The neo-Epicurean teaching of which we are all aware, the capital-*E* Evolutionism that has produced a metaphysical inflation from a proven hypothesis about the physical world to a naturalistic worldview—this modernist teaching *has exposed a flank that perhaps needed exposing.* When attacks on the faith come along, they routinely attack at the point where the church had left itself unguarded. Reformations have come when the church, to respond to a new challenge, has not been content to repeat slogans from the past, however true, but has engaged in urgent and fresh searching of scripture. This does not lead, as worried systematicians sometimes fear, to the erratic pseudopapacy of the quirky biblical scholars, but rather to biblical scholarship as the humble yet powerful task of serving the whole Body of the Messiah with the fresh insight it now needs. And I firmly believe that part of this task in our generation is to alert the church to the theme of the kingdom.

The central message of the Bible is not simply that we are sinners, but through Jesus God is rescuing us from this sinful world so that we can be with him in heaven. That's part of it, but it's not the whole biblical story. The Bible is not about the rescue of humans *from* the world but about the rescue of humans *for* the world, and indeed God's rescue of the world *by means of* those rescued humans. This is where we turn toward Genesis 1, toward a fresh reading of image and temple.

One key point to note as we turn to the second half of these reflections: it simply won't do to check the boxes of the traditional dogmas. Yes, Jesus was and is fully divine and fully human. But the *point* of his divinity in the Gospels is that in him and *as* him the living God is becoming king. And the point of his humanity in the Gospels is that, in him and as him, human beings are at last taking up again their God-given vocation of being the royal priesthood through which God brings his wise, redemptive ordering to the garden. And yes, the good news is good news of salvation. But in the Bible we are saved not simply so we can go to heaven and enjoy fellowship with God but so that we can be his truly human royal priesthood in his world. "You were slaughtered," sings the great crowd in Revelation 5 in praise of the Lamb, "and with your own blood you purchased a people for God, . . . *and made them a kingdom and priests to serve our God, and they will reign on the earth.*" And here it comes: if only we could get this picture straight, all our discussions about

Adam, about origins, about the meaning of Genesis 1–3, would appear in a quite different light.

Paul's Writings on Adam

At the center of the puzzle, of course, stands Paul's reference to Adam in Romans 5 and 1 Corinthians 15. The Romans reference in particular has caused people to say not simply that we have to believe in Adam and Eve because a literal reading of Genesis says so but also that we have to believe in a historical single Adam because Paul says so, and that the way Paul says it means that at the heart of his soteriology there is a contrast between Adam and Christ, and if you take away the first half, the second half is meaningless. In reply, I have two points to make. First, Paul's exposition of Adam in these passages is explicitly in the service not of a traditional soteriology but of the kingdom of God. Second, there is a close parallel between the biblical vocation of Adam in Genesis and the biblical vocation of Israel, and when we explore this we may find fresh ways through to the heart of our puzzles.

First, then, Adam and the kingdom of God. Despite many generations in which Romans has been read simply as a book about how we get saved, that is not the ultimate point even of chapters 1–8 or 5–8. The great climax of Romans 1–8 is the renewal of all creation, in 8:17–26, where Jesus as Messiah, with a reference to Psalm 2, is given as his inheritance the uttermost parts of the world. For Paul it's clear: *the whole world is now God's holy land.* That's what scripture prophesied, and that's what has been achieved in Jesus the Messiah. But this inheritance is shared with all Jesus's people, and this happens ultimately through their resurrection. "Creation itself," declares Paul, "would be freed from its slavery to decay, to enjoy the freedom that comes when God's children are glorified."

Now, please note, he doesn't mean that creation will *share* the glory, as some translations misleadingly suggest. Paul is working with Psalm 8 as well as Psalm 2, and in Psalm 8, exactly as in Genesis 1, humans are given glory and dominion over the world. Here is the problem to which Romans is the answer: not simply that we are sinful and need saving, but that our sinfulness has meant that God's project for the whole creation (that it should be run by obedient humans) was aborted, put on hold. And when we are saved, as Paul spells out, that is in order that the whole-creation project can at last get back on track. When humans are redeemed, creation gives a sigh of relief and says, "Thank goodness! About time you got yourselves sorted out!"

This, you see, is what Paul is really talking about in Romans 5:12–21, though of course there isn't time for a full exegesis here. I simply draw attention to verses 17 and 21. In verse 17, Paul surprises us. "If by the trespass of the one, death reigned through the one," he says, and we expect him to go on, "how much more will life reign through the one." But he doesn't. He says, "How much more will those who receive the abundance of grace, and of the gift of covenant membership, of "being in the right," reign in life through the one man Jesus the Messiah." Adam's sin meant not only that he died but that he no longer reigned over the world. God's creation was supposed to function through human stewardship, and instead it now produces thorns and thistles. Now humans are redeemed to get God's creation-project back on track, and the word for all that is *reigning,* "ruling," *basileuein* in Greek, in other words, "kingdom." Paul's Adam theology is also his kingdom theology, and the author of Genesis would have smiled in recognition. Verse 21 points, densely of course, in the same direction. Grace reigns *through righteousness* to the life of the Age to Come. God sets people right in order, through them, to set the world right. Justification by faith is the advance putting right of people in order to, through them, put the world right.

We see the same from a different angle in 1 Corinthians 15:20–28. Again Paul is working with the Psalms, in this case 110 and again 8. His point is that Jesus is already enthroned, already king, already reigning. In other words, *he is at last where Adam was supposed to be.* There is at last an obedient human at the helm of the universe. Of course, this is part of Paul's now-and-not-yet theology; Jesus is already reigning, but one day the last enemy will finally be overcome, namely death. Paul is working closely with Genesis 1, 2, and 3, right across 1 Corinthians 15, and one could write a whole book about the way that works out. Basic to his exposition of Genesis is this point: that God put his wonderful world into human hands; that the human hands messed up the project; and that the human hands of Jesus the Messiah have now picked it up, sorted it out, and got it back on track. What I really want to say about Paul and Adam is that it won't do simply to go to Paul and say, "There you are, he believes in Adam, and that proves our literalistic reading of Genesis." What it shows up is precisely the *failure* of the tradition to read either Paul or Genesis, because Paul's whole point is to pick up from Genesis the notion of *the calling of Adam* and to show that it is fulfilled in the Messiah. Unless we put that in the middle, we are not being obedient to the authority of these central scriptural texts.

This sends me back to Genesis, then, to look at the calling of Adam. The notion of the image doesn't refer to a particular spiritual endowment, a secret property that humans possess somewhere in their genetic makeup, something that might be found by a scientific observation of humans as opposed to chimps or dolphins. The image is a *vocation,* a calling. It is the call to be *an angled mirror,* reflecting God's wise order into the world and the praises of all creation back to the creator. That is what it means to be the royal priesthood: looking after God's world is the royal bit; summing up creation's praise is the priestly bit. And the image is, of course, the final thing that is put into the temple so that the god can be present to his people through the image and his people can worship him in that image. One of the great gains of biblical scholarship this last generation, not least because of our new understanding of first-century Judaism, is our realization that the temple was central to the Jewish worldview.

But here is the problem. We have seen the goal of it all as humans being rescued so that we could have fellowship with God, but the Bible sees the goal as humans being rescued so that we could sum up the praises of all creation and look after that creation as God's wise stewards. Genesis, the Gospels, Romans, and Revelation all insist that the problem goes like this: human sin has blocked God's purpose for creation, but God hasn't gone back on his creational purpose, which was and is to work in his creation through human beings, his image bearers. He hasn't gone back on the plan; through his true image bearer, Jesus the Messiah, he has rescued humans from sin and death in order to reinscribe his original purposes, which include the extension of sacred space into all creation, until the earth is indeed filled with God's knowledge and glory as the waters cover the sea. God will be present in and with his whole creation; the whole creation will be like a glorious extension of the tabernacle in the wilderness or the temple in Jerusalem.

This is the point where I sense a strong parallel with the calling and vocation of the ancient people of Israel, and this is the point where we might just cast a glimmer of fresh light back on Adam and the question of origins. Genesis itself makes a parallel: The command to Adam, to be fruitful and multiply, turns into the promise to Abraham, that God will bless him, make him fruitful, and multiply him. Instead of the original paradise, where God is present with his people, Israel is promised the land and eventually given the temple as the place of God's presence. But the point is this: Israel, a small, strange nomadic people in an obscure part of the world, is chosen to be the promise bearer: "In your descendants all the families of the earth shall be blessed." Israel is to be a royal priesthood (Exodus 19). Israel is to be the light of the nations (Isaiah 42, 49). *Israel is chosen out of the rest of the world to be God's strange means of rescuing the human race and so getting the creational project back on track.* And God chooses Israel knowing full well that, in Paul's language, Israel too is in Adam; the people who bear the solution are themselves part of the problem.

That is the clue to the hardest bits of Paul's theology, for instance the problem of the Law. That's for another time. But watch closely. Israel is chosen to fulfill this divine purpose; Israel is placed in the Holy Land, the garden of God's delight, and warned that if they don't keep Torah they will be expelled, sent off into exile. It will look as though the whole project has been aborted. That is the horrible problem faced not only in the exile but in the so-called postexilic period. And it is that complex problem which the New Testament sees being dealt with, gloriously resolved, in Israel's messiah, Jesus the Lord, and his death and resurrection. He has dealt with exile, and now the whole world is God's holy land, with Jesus and his people as the light of the world.

Instead of coming forward from the calling of Israel, come back to the calling of Adam, which is so closely parallel. I do not know when Genesis reached its final form. Some still want to associate it with Moses; others insist it was at least edited during the exile. But whatever view you take about that, certainly the Jews of the Second Temple period would have no difficulty in decoding the story of Adam as an earlier version of their own story: placed in the garden, given a commission to look after it; the garden being the place where God wanted to be at rest, to exercise his sovereign rule; the people warned about keeping the commandment, warned in particular that breaking it would mean death, breaking it, and being exiled. It all sounds very, very familiar.

And it leads me to my proposal: that just as God chose Israel from the rest of humankind for a special, strange, demanding vocation, so perhaps what Genesis is telling us is that *God chose one pair from the rest of early hominids for a special, strange, demanding vocation.* This pair (call them Adam and Eve if you like) were to be the representatives of the whole human race, the ones in whom God's purpose to make the whole world a place of delight and joy and order, eventually colonizing the whole creation, was to be taken forward. God the creator put into their hands the fragile task of being his image bearers. If they fail, they will bring the whole purpose for the wider creation, including all the nonchosen hominids, down with them. They are supposed to be the life bringers, and if they fail in their task the death that is already endemic in the world as it is will engulf them as well.

This, perhaps, is a way of reading the warning of Genesis 2: in the day you eat of it *you too will die*. Not that death, the decay and dissolution of plants, animals, and hominids wasn't a reality already; but you, Adam and Eve, are chosen to be the people through whom God's life-giving reflection will be imaged into the world, and if you choose to worship and serve the creation rather than the creator you will merely reflect death back to death, and will share that death yourself. I do not know whether this is exactly what Genesis meant or what Paul meant. But the close and (to a Jewish reader) rather obvious parallel between the vocation of Israel and the vocation of Adam leads me in that direction.

One might perhaps sum it up like this. The problem of Israel is that it is called to be God's means of rescuing the world, but Israel is part of the problem of Adam to which she is supposed to be providing the solution. In a similar way—not exactly parallel but similar—Adam and Eve are chosen to take the creator's purposes forward to a new dimension of life. But if they fail—if they abdicate their image-bearing vocation and follow the siren call of the elements of chaos still within creation—they will come to share the entropy that has so far been creation's lot. They do, and they do.

All this projects us forward toward a full and rich Christology, but not simply of Jesus as both divine and human—that's a given, but it's only a shorthand, a signpost. Look at Paul's language: Jesus is the beginning, the firstfruits, the true Image, the Temple in whom all God's fullness was pleased to dwell. He is Israel's

Messiah, who fulfills Israel's obedience on the cross and thereby rescues both Israel and the whole human race. He does for Israel what Israel couldn't do for itself, and thereby does for humans what Israel was supposed to do for them, *and thereby launches God's project of new creation, the new world over which he already reigns as king.* This is the great narrative, and we need to learn to tell it.

And here we stumble upon an interesting point: that just as the biologists and philosophers have pointed toward the complex notion of altruism as something that might just be a signpost away from the closed continuum of selfish genes, so in the Christian message we have the cross, not just as an act of altruism—*altruism* is after all a thin, bloodless word, a parody of the reality—but the supreme act of love. "The son of God loved me and gave himself for me," wrote Paul. "He had always loved his own people in the world," wrote John; "now he loved them right through to the end." The cross is, and Jesus always said it was, the subversion of all human power systems. The cross is the central thing that demonstrates the impossibility of the metaphysically inflated Evolution-with-a-capital-*E*. The weakness of God is stronger than human strength. And it leads, as Jesus said it would lead, to a life of following him, which would itself be about taking up the cross and so finding life, about the meek inheriting the earth. The point of it all, once more, is *vocational:* if we can study Genesis and human origins without hearing *the call to be an image-bearing human being renewed in Jesus,* we are massively missing the point, perhaps pursuing our own dream of an otherworldly salvation that merely colludes with the forces of evil, as gnosticism always does.

SOMEONE ONCE ASKED ME how I could preach all this. Well, it may require some adjustment not only in content but also in style. That's always likely to be the challenge for a good preacher. But believe me, this theme will preach. It's the greatest story ever told, and it will draw all our stories up into it. Yes, many humans, though not all, are deeply aware of problems in their own lives, of pains and fears and sorrows and deep-rooted puzzles, and that may well bring them to the foot of the cross. But the message ought never to be simply about "me and my salvation." It ought to be about God and God's kingdom. That's what Jesus announced, and so should we. The full good news is that in Jesus God has become king of the world; we look out at the world and see it in a terrible mess, and we are aware in our bones that we want to do something about it. But our sin, our greed, our pride, our arrogance get in the way, and we rush off and try to do it with our own strength and (worse) in our own way, like Moses trying to liberate Israel from Egypt by Egyptian means. He first needed liberating himself.

We humans know in our bones that we are called to bring God's wise order into the world, but for that to become a reality we need ourselves to be rescued from the same problem the rest of the world is in. We are rescued by the blood of the Lamb to be a royal priesthood, and the way that works, according to the New Testament, is the same way it worked for Jesus: taking up the cross as a suffering but joyful witness. There is plenty there for years and years of sermons. They won't be comfortable either to preach or to listen to. But that was never the point, was it? Leave the dead to bury their dead, said Jesus. You go and tell people that this is the time when God is becoming king.

3

Can a Scientist Believe
in the Resurrection?

A QUESTION BEGINNING "Can a scientist . . ." is a dangerous one for me to address. I make no claim to know anything about science. I did precisely one year of physics and chemistry at school, and since I knew before I began that I was going to give them up to concentrate on classics, I did as little work as I could without actually entering a penal zone. In fact, my chemistry report in summer 1963 said, "He has maintained his position"—which was twenty-fourth out of twenty-four—"with occasional signs of interest now and then." I did, however, love mathematics, with its elegance and harmonies. It was the subject I was most sorry to give up, but that's another story.

Of course, it is possible to give a short and trivial answer to the question "Can a scientist believe in the resurrection?" rather like the man who, when asked whether he believed in infant baptism, replied, "Sure! I've seen it done!" That, of course, exposes one of the problems with the phrase *believe in:* does it mean "believe that it *can* be done" or "believe that it *should* be done"? And there are other possibilities too, as we will see. Similarly, to the question

"Can a scientist believe in the resurrection?" one might simply reply, "Sure! I've seen it done!" I know plenty of scientists who firmly and avowedly believe in the resurrection, and some indeed who have given a solid and coherent account of why they do so. I salute them but do not intend to engage with the different ways in which they have presented their case. I want, rather, to explore the fault lines, if that's the right expression, between different ways of knowing, particularly between what we may loosely call scientific knowing and historical knowing, and between both of these and those other modes of knowing to which we give, very loosely, the names faith, hope, and love.

My case, you will not be surprised to learn, is that these ways of knowing overlap and interlock much more than we usually suppose. Certainly much more than a certain kind of rhetoric would try to persuade us: it has been a feature of the last two hundred years to invoke a kind of pan-Enlightenment thesis, namely that the methods and results of modern science have delivered us from the dark superstitions of the past, sometimes designated "medieval," so that everything that happened before, say, 1750, with a few golden exceptions, was ignorance and guesswork and everything since then has been an upward path toward the light. I am sometimes accused of being anti-Enlightenment, and there is a grain of truth in that because I do think that postmodernity has got some important points to make; but I want to assure you that I have no wish to return to pre-Enlightenment dentistry, sanitation, or travel, to look no farther. I merely note that there are obvious ambiguities as well as massive gains. The movement that gave us penicillin also gave us Hiroshima. Somehow, as most admit and I suspect all know in their bones, science in the strict sense can never be enough—enough, that is, for a full and flourishing human life in all its dimensions.

But the question then turns on the word *believe,* and here too are puzzles to explore. Plato declared that belief was a kind of second-rate knowing, more or less halfway between knowing and not knowing, so that the objects of belief possessed a kind of intermediate ontology, halfway between existence and nonexistence. This way of thinking has colored popular usage, so that when we say, "I believe it's raining," we are cushioning ourselves against the possibility that we might be wrong, whereas when we say, "I know it's raining," we are open to straightforward contradiction. But this usage has slid, over the last centuries, to the point where, with a kind of implicit positivism, we use *know* and *knowledge* for things we think we can in some sense prove, and *believe* and its cognates for things that we perceive as degenerating into mere private opinion without much purchase on the wider world.

And the Christian claim was from the beginning that Jesus's resurrection was a question not of the internal mental and spiritual states of his followers a few days after his crucifixion but of something that had happened in the real, public world, leaving among its physical mementos not only an empty tomb but a broken loaf at Emmaus and footprints in the sand by the lake, and leaving his followers with a lot of explaining to do but with a transformed worldview that is only explicable on the assumption that something really did happen, even though it stretched their existing worldviews to the breaking point. More of that shortly. What we now have to do is to examine this early Christian claim more thoroughly, to ask what can be said about it historically, and to inquire particularly what sort of knowing or believing we are talking about when we ask whether a scientist can believe that which, it seems, the word "resurrection" actually refers to.

Different Types of Knowing

First, some reflections—unsystematic musings, really—on the types of knowing. I assume that when we ask whether a scientist can believe something, we are asking a two-level question. First, we are asking about what sort of things the scientific method can explore, and how it can know or believe certain things. Second, we are asking about the kind of commitment someone wedded to scientific knowing is expected to have in all other areas of his or her life.

Is a scientist, for example, expected to have a scientific approach to listening to music? To watching a football game? To falling in love? The question assumes, I think, that resurrection, and perhaps particularly the resurrection of Jesus, is something that might be expected to impinge on the scientist's area of concern, somewhat as if one were to say, "Can a scientist believe that the sun could rise twice in a day?" or "Can a scientist believe that a moth could fly to the moon?" (I did actually watch the sun *set* twice in a day; I took off from Aberdeen on a winter afternoon shortly after sunset, and the sun rose again as we climbed, only then to set, gloriously, a second time shortly afterward. That was, of course, cheating.) This is different, in other words, from saying, "Can a scientist believe that Schubert's music is beautiful?" or "Can a scientist believe that her husband loves her?" There are those, of course, who, by redefining the resurrection to make it simply a spiritual experience in the inner hearts and minds of the disciples, have pulled the question toward the latter pair and away from the former. But that is ruled out by what, as we will see, all first-century users of the language of resurrection meant by the word. *Resurrection* in the first century meant people who were physically thoroughly dead becoming physically thoroughly alive again, not simply surviving or entering a purely spiritual world, whatever that might be. And resurrection therefore necessarily impinges on the public world.

But it is the public world not of the natural scientist but of the historian. To put it crudely, and again without all the necessary footnotes and nuances, science studies the repeatable, while history studies the unrepeatable. Caesar only crossed the Rubicon once, and if he'd crossed it again it would have meant something different the second time. There was, and could be, only one first landing on the moon. The fall of the second Jerusalem Temple took place in AD 70 and never happened again. Historians don't, of course, see this as a problem and are usually not shy about declaring that these events certainly took place even though we can't repeat them in the laboratory. But when people say, "But that can't have happened, because we know that *that sort of thing* doesn't actually happen," they are appealing to a kind of would-be scientific principle of history, namely the principle of *analogy*.

The problem with analogy is that it never quite gets you far enough, precisely because history is full of unlikely things that happened once and once only, so that the analogies are often at best partial, and are dependent anyway on the retort "who says?" to the objection about some kinds of things not normally happening. And indeed, in the case in point, we should note as an obvious but often overlooked point the fact that the early Christians did *not* think that Jesus's resurrection was one instance of something that happened from time to time elsewhere. Granted, they saw it as the first, advance instance of something that would eventually happen to everyone else, but they didn't employ that future hope as an analogy from which to argue back that it had happened already in this one instance.

So how does the historian work when the evidence points toward things that we do not normally expect? The resurrection is such a prime example of this that it's hard to produce, at this metalevel, analogies for the question. But, sooner or later, questions of worldview begin to loom in the background, and the question of what kinds of material the historian will allow on stage is inevitably affected by the worldview within which he or she lives. And at that point we are back to the question of the scientist who, faced with the thoroughly repeatable experiment of what happens to dead bodies, what it seems has always happened, and what seems likely always to go on happening, declares that the evidence is so massive that it is impossible to believe in the resurrection without ceasing to be a scientist altogether.

This is the point at which we must switch tracks and go to the evidence itself. What can be said, within whatever can be called scientific historiography, about the proposition that Jesus of Nazareth was bodily raised from the dead?

The Surprising Character of Early Christian Hope

I have sketched elsewhere the map of ancient beliefs about life beyond the grave. Ancient paganism contains all kinds of theories, but whenever resurrection is mentioned, the answer is a firm negative: we know that doesn't happen. (This is worth stressing in today's context. One sometimes hears it said or implied that prior to the rise of modern science people believed in all kinds of odd things like resurrection, but that now, with two hundred years of scientific research on our side, we know that dead people stay dead. This is ridiculous. The evidence was massive and the conclusion universally drawn as much in the ancient world as they are today.)

Ancient Judaism, however, is rooted in the belief that God is the creator of the world, and he will one day put the world to rights; this double belief, when worked out and thought through not least in times of persecution and martyrdom, produced by the time of Jesus a majority belief in ultimate bodily resurrection. The early Christian belief in hope beyond death thus belongs demonstrably on the Jewish, not the pagan, map. But the foundation of my argument for what happened at Easter is the reflection that this Jewish hope has undergone remarkable modifications or mutations within early Christianity, which can be plotted consistently right across the first two centuries. And these mutations are so striking, in an area of human experience where societies tend to be very conservative, that they force the historian, not least the would-be scientific historian, to ask, "Why did they occur?"

The mutations occur within a strictly Jewish context. The

early Christians held firmly, like most of their Jewish contemporaries, to a two-step belief about the future: first, death and whatever lies immediately beyond; second, a new bodily existence in a newly remade world. *Resurrection* is not a fancy word for "life after death"; it denotes life *after* "life after death." (There is much more to be said on this topic, but not here; for details, see both *The Resurrection of the Son of God* and *Surprised by Hope.*) There is nothing remotely like this in paganism. This belief is as Jewish as you can get. But within this Jewish belief there are seven early Christian mutations, each of which crops us in writers as diverse as Paul and John the Seer, as Luke and Justin Martyr, as Matthew and Irenaeus.

The first modification is that there is virtually no spectrum of belief on this subject within early Christianity. The early Christians came from many strands within Judaism and from widely differing backgrounds within paganism, hence from circles that must have held very different beliefs about life beyond death. But they all modified that belief to focus on one point on the spectrum. Christianity looks, to this extent, like a variety of Pharisaic Judaism. There is no trace of a Sadducean view, or of that of Philo. For almost all of the first two centuries, resurrection in the traditional sense holds not only center stage in Christian belief about the ultimate future but the whole stage.

This leads to the second mutation. In Second Temple Judaism, resurrection is important but not that important. Lots of lengthy works never mention the question, let alone this answer. It is still difficult to be sure what the Dead Sea Scrolls thought on the topic. But in early Christianity, resurrection has moved from the circumference to the center. You can't imagine Paul's thought without it. You shouldn't imagine John's thought without it, though some have tried. Take away the stories of Jesus's birth, and all you lose is four chapters of the Gospels. Take away the resurrection and you lose the entire New Testament, and most of the second-century fathers as well.

The third mutation has to do with what precisely resurrection *means*. In Judaism it is usually left vague as to what sort of body the resurrected will possess; some see it as a resuscitated but basically identical body, while others think of it as a shining star. But from the start the early Christians believed that the resurrection body, though it would certainly *be* a body in the sense of a physical object, would be a transformed body, a body whose material, created from the old material, would have new properties. That is what Paul means by the "spiritual body": not a body made out of nonphysical spirit, but a physical body animated by the Spirit, a Spirit-*driven* body if you like: still what we would call physical but differently animated. And the point about this body is that, whereas the present flesh and blood are corruptible, doomed to decay and die, the new body will be incorruptible. One Corinthians 15, one of Paul's longest sustained discussions and the climax of the whole letter, is about the creator God remaking the creation—not abandoning it, as Platonists of all sorts, including the gnostics, would have wanted.

The fourth surprising mutation in early Christian resurrection belief is that the resurrection, as an event, has split into two. No first-century Jew, prior to Easter, expected the resurrection to be anything other than a large-scale event happening to all God's people, or perhaps to the entire human race, at the very end. There were, of course, other Jewish movements that held some kind of inaugurated eschatology. But we never find outside Christianity what becomes a central feature within it: the belief that the resurrection itself has happened to one person in the middle of history, anticipating and guaranteeing the final resurrection of his people at the end of history.

I am indebted to John Dominic Crossan for highlighting what I now list as the fifth mutation in Jewish resurrection belief. In a public debate, Crossan spoke of "collaborative eschatology." Because the early Christians believed that resurrection had begun with Jesus and would be completed in the great final resurrection

on the last day, they believed also that God had called them to
work with him, in the power of the Spirit, to implement the
achievement of Jesus and thereby anticipate the final resurrection,
in personal and political life, in mission and holiness. If Jesus, the
Messiah, was God's future arriving in person in the present, then
those who belonged to Jesus and followed him in the power of his
spirit were charged with transforming the present, as far as they
were able, in the light of that future.

The sixth mutation in Jewish belief is the new metaphorical use
of *resurrection*. I have written about that elsewhere. Basically, in the
Old Testament *resurrection* functions once, famously, as a metaphor
for return from exile (Ezekiel 37). In the New Testament, that
sense has disappeared, and a new metaphorical use has emerged,
with *resurrection* used in relation to baptism and holiness (Romans 6; Colossians 2–3), though without, importantly, affecting
the concrete referent of a future resurrection (Romans 8).

The seventh and final mutation within Jewish resurrection belief is its association with messiahship. Nobody in Judaism had
expected the Messiah to die, and therefore naturally nobody had
imagined the Messiah rising from the dead. This leads us to the
remarkable modification not just of resurrection belief but of messianic belief. Where messianic speculations existed (again, by no
means all Jewish texts spoke of a messiah, but the notion became
central in early Christianity), the Messiah was supposed to fight
God's victorious battle against the wicked pagans, to rebuild or
cleanse the temple, and to bring God's justice to the world. Jesus,
it appeared, had done none of these things. No Jew with any idea
of how the language of messiahship worked at the time could
have possibly imagined, after his crucifixion, that Jesus of Nazareth was indeed the Lord's anointed. But from very early on, as
witnessed by what may be pre-Pauline fragments of early creedal
belief, such as Romans 1:3–4, Christians affirmed that Jesus was
indeed the Messiah, precisely because of his resurrection.

We note at this point, as an important aside, how impossible it is to account for the early Christian belief in Jesus as Messiah without the resurrection. We know of several other Jewish movements, messianic movements, prophetic movements, during the one or two centuries on either side of Jesus's public career. Routinely they ended with the violent death of the central figure. Members of the movement (always supposing they got away with their own skins) then faced a choice: either give up the struggle or find a new messiah. Had the early Christians wanted to go the latter route, they had an obvious candidate: James, the Lord's brother, a great and devout teacher, the central figure in the early Jerusalem church. But nobody ever imagined that James might be the Messiah.

This rules out the revisionist positions on Jesus's resurrection that have been offered by so many writers in recent years. Suppose we go to Rome in AD 70 and there witness the flogging and execution of Simon bar Giora, the supposed king of the Jews, brought back in Titus's triumph. Suppose we imagine a few Jewish revolutionaries three days or three weeks later.

The first revolutionary says, "You know, I think Simon really was the Messiah—and he still is!"

The others would be puzzled. "Of course he isn't; the Romans got him, as they always do. If you want a messiah, you'd better find another one."

"Ah," says the first, "but I believe he's been raised from the dead."

"What d'you mean?" his friends ask. "He's dead and buried."

"Oh no," replies the first, "I believe he's been exalted to heaven."

The others look puzzled. "All the righteous martyrs are with God; everybody knows that. Their souls are in God's hand, but that doesn't mean they've *already* been raised from the dead. Anyway, the resurrection will happen to us all at the end of time, not to one person in the middle of continuing history."

"No," replies the first, anticipating the position of twentieth-century existentialist theology, "you don't understand. I've had a strong sense of God's love surrounding me. I have felt God forgiving me—forgiving us all. I've had my heart strangely warmed. What's more, last night, I saw Simon; he was there with me. . . ."

The others interrupt, now angry. "We can all have visions. Plenty of people dream about recently dead friends. Sometimes it's very vivid. That doesn't mean they've been raised from the dead. It certainly doesn't mean that one of them is the Messiah. And if your heart has been warmed, then for goodness's sake sing a psalm, but don't make wild claims about Simon."

That is what they would have said to anyone offering the kind of statement that, according to the revisionists, someone must have come up with as the beginning of the idea of Jesus's resurrection. But this solution isn't just incredible; it's impossible. Had anyone said what the revisionists suggest, some such conversation as the above would have ensued. A little bit of disciplined historical imagination is all it takes to blow away enormous piles of so-called historical criticism.

What is more—to round off this final mutation from within the Jewish belief—because of the early Christian belief in Jesus as Messiah, we find the development of the very early belief that Jesus is Lord and that therefore Caesar is not. This is a whole other topic for another occasion. Death is the last weapon of the tyrant; the point of the resurrection, despite much misunderstanding, is that death has been defeated.

We have thus noted seven major mutations within Jewish resurrection belief, each of which became central to early Christianity. The belief in resurrection remains emphatically on the map of first-century Judaism rather than paganism; but, from within the Jewish theology of monotheism, election, and eschatology, it has opened up a whole new way of seeing history, hope, and hermeneutics. *And this demands a historical explanation.* Why did the early Christians modify Jewish resurrection language in these seven ways and do it with such consistency? When we ask them, they reply that they have done it because of what they believe happened to Jesus on the third day after he died. This forces us to ask: What then must we say about the very strange stories they tell about that first day?

The First-Century Stories of Easter

When we plunge into the stories of the first Easter Day—the accounts we find in the closing chapters of the four canonical Gospels—we find that, notoriously, the accounts do not fit snugly together. How many women went to the tomb, and how many angels or men did they meet there? Did the disciples meet Jesus in Jerusalem or Galilee or both? And so on. At this point I would like to invoke the splendid story of what happened in October 1946 when Karl Popper gave a paper at Wittgenstein's seminar in King's College, as written up in the book *Wittgenstein's Poker.* Several highly intelligent men—men who would modestly have agreed that they were among the most intelligent men in the world at the time—were in the room as Wittgenstein brandished a poker about and then left abruptly, but none of them could quite agree afterward as to what precisely had happened. But, as with Cambridge in 1946, so with Jerusalem in AD 30 (or whenever it was): surface discrepancies do not mean that nothing happened. Indeed, they are a reasonable indication that something remarkable happened.

As part of the larger argument that I have advanced elsewhere, I here draw attention to four strange features shared by the accounts in the four canonical Gospels. These features, I suggest, compel us to take them seriously as very early accounts and not, as is often suggested, later inventions.

First, we note the strange absence of the Old Testament in the stories. Up to this point, all four evangelists have drawn heavily upon biblical quotation, allusion, and echo. But the resurrection narratives are almost entirely innocent of them. This is the more remarkable, in that from as early as Paul the common credal formula declared that the resurrection too was "in accordance with the Bible," and Paul and the others ransack psalms and prophets to find texts that will explain what has just happened and to set it within, and as the climax to, the long story of God and Israel. Why do the Gospel resurrection narratives not do the same?

We could say, of course, that whoever wrote the stories in the form we now have them had gone through, cunningly, and taken material out *to make them look as if they were very old,* rather like someone deliberately taking all the electrical fittings out of a house to make it look like it might have a century or more ago. That might be marginally plausible if we had just one account, or if the four accounts were obviously derived from one another. We don't, and they aren't. You either have to imagine four very different writers each deciding to write up an Easter narrative based on the theology of the early church but with the biblical echoes taken out; or you have to say, which is infinitely more probable, that the stories, even though written down a lot later, go back to an extremely early oral tradition that had been formed and set firmly in the memory of different storytellers before there was time for biblical reflection.

The second strange feature of the stories is better known: the presence of the women as the principal witnesses. Whether we like it or not, women were not regarded as credible witnesses in the ancient world. Nobody would have made them up. Had the tradition started in the male-only form we find in 1 Corinthians 15, it would never have developed—and in such different ways—into the female-first stories we find in the Gospels. The Gospels must embody the earliest storytelling, and 1 Corinthians 15 a later revision.

The third strange feature is the portrait of Jesus himself. If, as many revisionists have proposed, the Gospel stories developed either from people mulling over the scriptures following Jesus's death or a new experience of inner illumination, you would expect to find the risen Jesus shining like a star. That's what Daniel says will happen. We have such a story in the Transfiguration. But none of the Gospels say this about Jesus at Easter. Indeed, he appears as a human being with a body that in some ways is quite normal, and he can be mistaken for a gardener or a fellow traveler on the road. Yet the stories also contain mysterious but definite signs that this body has been transformed. It is clearly physical, using up (so to speak) the matter of the crucified body; hence the empty tomb. But equally, it comes and goes through locked doors; it is not always recognized; and in the end it disappears into God's space, that is, heaven, through the thin curtain that in much Jewish thought separates God's space from human space.

This kind of account is without precedent, biblical or otherwise, and it looks as if the writers knew it. And this rules out the old idea that Luke's and John's accounts, which are the most apparently physical, were written late in the first century in an attempt to combat Docetism (the view that Jesus wasn't a real human being but only seemed to be so). If Luke and John were combating Docetism, they would never have said that the risen Jesus appeared and disappeared through locked doors, sometimes being recognized, sometimes not, and finally ascended into heaven.

Let me just add here as a footnote: In a review of my book *The Resurrection of the Son of God,* in *Scottish Journal of Theology* no less, Michael Welker, though saying some very flattering things as well, accuses me of basically saying that Jesus was "resuscitated." I am very puzzled by this since I took pains to make it clear that there is all the difference in the world between returning from the dead into the same kind of corruptible body, which will have to die again, and going through death and out the other side into a new type of physicality. This new physicality will, it seems, have two properties in particular: first, it was and presumably still is equally at home in heaven and earth; second, though in our sense solid and physical, it was and is no longer corruptible, not capable of decay or death.

The fourth strange feature of the resurrection accounts is the entire absence of mention of the future Christian hope. Almost everywhere else in the New Testament, the resurrection of Jesus is spoken of in connection with the final hope that those who belong to Jesus will one day be raised as he has been, with the note that this must be anticipated in the present through baptism and behavior. Insofar as the event is interpreted, it has a very this-worldly, present-age meaning: Jesus is raised, so he is the Messiah, the world's true lord. The long-awaited new creation has begun—and we therefore have a job to do, to act as Jesus's heralds to the entire world. Once again, had the stories been invented toward the end of the first century, this interpretation would certainly have included a mention of the final resurrection of all God's people.

What do we conclude from all this? That the stories, though lightly edited and written down later, are basically very, very early. They are not, as has so often been suggested, legends written much later to give a pseudohistorical basis for what had been essentially a private experience. And when we ask how such stories could have come into existence, the obvious answer all the early Christians give is that, though it was hard to describe at the time and remains mind-boggling thereafter, something like this is what happened. And it is now time to ask, at last: What can the historian today say about all this? And then, what can the scientist say about it?

What History and Science Have to Say About Easter

The only way we can explain the phenomena we have been examining is by proposing a two-pronged hypothesis: first, Jesus's tomb really was empty; second, the disciples really did encounter him in ways which convinced them that he was not simply a ghost or hallucination. A brief word about each.

For the disciples to see, or think they saw, someone they took to be Jesus would not by itself have generated the stories we have. Everyone in the ancient world (like many today) knew that people sometimes had strange experiences involving encounters with the dead, particularly the recently dead. However many such visions they had had, they wouldn't have said Jesus was raised from the dead; they weren't expecting such a resurrection. In any case, Jesus's burial was a standard primary burial that would require a secondary burial in an ossuary at some later point. Someone would have had to go and collect Jesus's bones, fold them up, and store them. Nobody in the Jewish world would have spoken of such a person being already raised from the dead. Without the empty tomb, they would have been as quick to say "hallucination" as we would.

Equally, an empty tomb by itself proves almost nothing. It might (as many have suggested) have been the wrong one, though a quick check would have sorted that out. The soldiers, the gardeners, the chief priests, other disciples, or someone else might have taken away the body. That was the conclusion Mary drew in John's Gospel and the story the Jewish leaders put about in Matthew's. Unless the finding of the empty tomb had been accompanied by sightings of and meetings with the risen Jesus, that is the kind of conclusion they would all have drawn. The meetings on the one hand and the empty tomb on the other are therefore both necessary if we are to explain the rise of the belief and the writing of the stories as we have them. Neither by itself would be sufficient; put them together, though, and they provide a complete and coherent explanation for the early Christian belief.

All this brings us face-to-face with the ultimate question. The empty tomb and the meetings with Jesus are, in combination, the only possible explanation for the stories and beliefs that grew up so quickly among his followers. How, in turn, do we explain *them*? What can the historian say? What can the scientist say?

In any other historical inquiry, the answer would be so obvious that it would hardly need saying: the best explanation is that it happened that way. Here, of course, it is so shocking, so earth shattering, that we rightly pause before leaping into the unknown. And here, indeed, as some skeptical friends have cheerfully pointed out to me, it is always possible for someone to follow the argument so far and to say simply, "I don't have a good explanation for what happened to cause the empty tomb and the appearances, but I choose to maintain my belief that dead people don't rise and therefore conclude that something else must have happened even though we can't tell what it was." That is fine; I respect that position, but I simply note that it is indeed then a matter of choice, *not* a matter of saying that something called scientific historiography forces us to take that route.

But at this moment in the argument, all the signposts point in one direction. I have examined elsewhere all the alternative explanations, ancient and modern, for the rise of the early church, and I have to say that far and away the best historical explanation is that Jesus of Nazareth, having been thoroughly dead and buried, really was raised to life on the third day with a new *kind* of physical body, which left an empty tomb behind it because it had used up the material of Jesus's original body and possessed new properties that nobody had expected or imagined but which generated significant mutations in the thinking of those who encountered it. If something like this happened, it would perfectly explain why Christianity began and why it took the shape it did.

But this is where I want to heed carefully the warnings of those theologians who have cautioned against any attempt to stand on the ground of rationalism and attempt to prove, in some mathematical fashion, something that, if it happened, ought itself to be regarded as the center not only of history but also of epistemology, not only of *what* we know but of *how* we know it. This is where the third element in knowing, the puzzling bits beyond science or history but still interacting with both, inevitably come into play.

I once imagined, to make this point, a fantasy scenario: a rich alumnus gives to a university a wonderful, glorious painting that simply won't fit any of the spaces available in the university. The painting is so magnificent that eventually the university decides to pull itself down and rebuild itself around this great and unexpected gift, discovering as it does so that all the best things about the university the way it was are enhanced within the new structure, and all the problems of which people had already been aware are thereby dealt with. And the key thing about that illustration, inadequate though it is, is that there must be some point at which the painting is received by the existing university, some epistemological overlap point to enable the college officers to make their momentous decision. The donor doesn't just come along, demolish the university unasked, present the painting, and then say, "Now figure out what to do."

My point is that the resurrection of Jesus, presenting itself as the obvious answer to the question "How do you explain the rise of early Christianity?" has that kind of purchase on serious historical inquiry *within the present world,* and therefore poses that kind of challenge to the larger worldview of both the historian and the scientist. (It isn't, in other words, like the kind of über-Barthian apologetic that simply says, "Here is the new world; get used to it, because we haven't got anything to say to you within your world." Nor is it like the rationalist apologetic, which offers a proof that not only begins *but also concludes* with the terms of the present creation, and therefore has to offer a supernatural account that concedes the split-level point it ought to be challenging.)

The challenge is in fact that of *new creation.* To put it at its most basic, the resurrection of Jesus offers itself, to the student of history or science no less than the Christian or the theologian, not as a very odd event within the world as it is, but as the utterly characteristic, prototypical, and foundational event within the world as it has begun to be. It is not an absurd event within the old world but the symbol and starting point of the new world. The claim advanced in Christianity is of that magnitude: that with Jesus of Nazareth there is not simply a new religious possibility, not simply a new ethic or a new way of salvation, but a new creation.

Now, that might seem to be an epistemological, as well as theological, preemptive strike. If there really is a new creation on the loose, the historian wouldn't have any analogies for it, and the scientist wouldn't be able to rank its characteristic events with other events that might otherwise have been open to inspection. What are we to do? No other explanations have been offered, in two thousand years of sneering skepticism against the Christian witness, that can satisfactorily account for how the tomb came to be empty, how the disciples came to see Jesus, and how their lives and worldviews were transformed. But history alone, certainly as conceived within the modern Western world and placed on the procrustean bed of science that (rightly) observes the world as it is, appears to leave us like the children of Israel shivering on the seashore. It can press the question to which Christian faith is the obvious answer. But if someone chooses to stay between the pharaoh of skepticism and the deep sea of faith, history cannot force them farther.

Everything then depends on the context within which the history is done. The most important decisions we make in life are not taken by post-Enlightenment left-brain rationality alone. I would not suggest that one can argue right up to the central truth of Christian faith by pure human reason building on simple observation of the world. Indeed, it should be obvious that one cannot. Equally, I would not suggest that historical investigation of this sort has therefore no part to play and all that is required is a leap of blind faith. God has given us minds to think; the question has been appropriately raised; Christianity appeals to history, and to history it must go. And the question of Jesus's resurrection, though it may in some senses burst the boundaries of history, also remains within them; that is precisely why it is so important, so disturbing, so life-and-death. We could cope—the world could cope—with a Jesus who ultimately remains a wonderful idea inside his disciples' minds and hearts. The world cannot cope with a Jesus who comes out of the tomb, who inaugurates God's new creation right in the middle of the old one.

That is why, for a complete approach to the question, we need to locate our study of history, and indeed of science, within a larger complex of human, personal, and corporate contexts, and this poses a challenge not only to the historian, not only to the scientist, but to all humans in whatever worldview they habitually live. The story of Thomas in John 20 serves as a parable for all of this. Thomas, like a good historian, wants to see and touch. Jesus presents himself to his sight and invites him to touch, but Thomas doesn't. He transcends the type of knowing he had intended to use and passes into a higher and richer one. Suddenly the new, giddying possibility appears before him: a new creation. Thomas takes a deep breath and brings history and faith together in a rush. "My Lord," he says, "and my God."

That is not an antihistorical statement, since the lord in question is precisely the one who is the climax of Israel's history and the launch of a new history, and once you grasp the resurrection you see that Israel's history is full of partial and preparatory analogies for this moment, so that the epistemological weight is borne not by the promise of ultimate resurrection and new creation alone but by the narrative of God's mighty actions in the past. Nor is it an antiscientific statement, since the world of new creation is precisely the world of new *creation* and as such open to and indeed eager for the work of human beings not to manipulate it with magic tricks, nor to be subservient to it as though the world of creation were itself divine, but to be its stewards; and stewards need to pay close, minute attention to that of which they are stewards, the better to serve it and enable it to attain its intended fruitfulness.

What I am suggesting is that faith in Jesus risen from the dead *transcends but includes* what we call history and what we call science. Faith of this sort is not blind belief that rejects all history and science. Nor is it simply—which would be much safer!—a belief that inhabits a totally different sphere, discontinuous from either, in a separate watertight compartment. Rather, this kind of faith, which is like all modes of knowledge defined by the nature of its object, is faith in the creator God, the God who has promised to put all things to rights at the end, the God who (as the sharp point where those two come together) has raised Jesus from the dead *within* history, leaving as I said evidence that demands an explanation from the scientist as well as anybody else. Insofar as I understand scientific method, when something turns up that doesn't fit the paradigm you're working with, one option at least, perhaps when all others have failed, is to change the paradigm, not to exclude everything you've known to that point but to include it within a larger whole. That is, if you like, the Thomas challenge.

If Thomas represents an epistemology of faith, which transcends but also includes historical and scientific knowing, we might suggest that Paul represents at this point an epistemology of hope. In 1 Corinthians 15, he sketches his argument that there will be a future resurrection, as part of God's new creation, the redemption of the entire cosmos as in Romans 8. Hope, for the Christian, is not wishful thinking or mere blind optimism. It is a mode of knowing, a mode within which new things are possible, options are not shut down, and new creation can happen. There is more to be said about this, but not here.

I want to finish with Peter. Epistemologies of faith and hope, both transcending but including historical and scientific knowing, point to an epistemology of love—an idea I first met in Bernard Lonergan, but which was hardly new with him. The story of John 21 sharpens it up. Peter, famously, has denied Jesus. He has chosen to live in the normal world, where the tyrants win in the end, and it's better to dissociate yourself from people who get on the wrong side of them. But now, with Easter, Peter is called to live in a new and different world; where Thomas is called to a new kind of faith and Paul to a radically renewed hope, Peter is called to a new kind of love. Here I go back to Wittgenstein once more, not this time for a poker but for a famous and haunting aphorism: "It is *love* that believes the resurrection." "Simon, son of John," says Jesus, "do you love me?" There is a whole world in that question, a world of personal invitation and challenge, of the remaking of a human being after disloyalty and disaster, of the refashioning of epistemology itself, the question of how we know things, to correspond to the new ontology, the question of what God's new world is like.

The reality that is the resurrection cannot simply be known from within the old world of decay and denial, tyrants and torture, disobedience and death. But that's the point. As I said, the resurrection is not, as it were, a highly peculiar event within the *present* world, though it is also that; it is the defining, central, prototypical event of the *new* creation, the world that is being born with Jesus. If we are even to glimpse this new world, let alone enter it, we will need a different kind of knowing, a knowing that involves us in new ways, an epistemology that draws from us not just the cool appraisal of detached quasi-scientific research but the whole-person engagement and involvement for which the best shorthand is "love," in the full Johannine sense of *agápē*. My sense from talking to scientific colleagues is that, though it's hard to describe, something like this is already at work when the scientist devotes him- or herself to the subject matter, so that the birth of new hypotheses seems to come about not so much through an abstract brain (a computer made of meat?) crunching data from elsewhere but more through a soft and mysterious symbiosis of knower and known, lover and beloved.

The skeptic will quickly suggest that this is, after all, a way of collapsing the truth of Easter once more into mere subjectivism. Not so. Just because it takes *agápē* to believe the resurrection, that doesn't mean all that happened was that Peter and the others felt their hearts strangely warmed. Precisely because it is *love* we are talking about, not lust, it must have a correlative reality in the world outside the lover. Love is the deepest mode of knowing, because it is love that, while completely engaging with reality other than itself, affirms and celebrates that other-than-self reality. This is the mode of knowing that is necessary if we are to live in the new public world, the world launched at Easter, the world in which Jesus is Lord and Caesar isn't.

That is why, although the historical arguments for Jesus's bodily resurrection are truly strong, we must never suppose that they will do more than bring people to the questions faced by Thomas and Peter, the questions of faith and love. We cannot use a supposedly objective historical epistemology as the ultimate ground for the truth of Easter. To do so would be like someone who lit a candle to see whether the sun had risen. What the candles of historical scholarship will do is show that the room has been disturbed, that it doesn't look like it did last night, and that would-be normal explanations for this won't do. Maybe, we think after the historical arguments have done their work, maybe morning has come and the world has woken up. But to find out whether this is so, we must take the risk and open the curtains to the rising sun. When we do so, we won't rely on candles anymore, not because we don't believe in evidence and argument, not because we don't believe in history or science, but because they will have been overtaken by the larger reality from which they borrow, to which they point, and in which they will find a new and larger home. All knowing is a gift from God, historical and scientific knowing no less than that of faith, hope, and love; but the greatest of these is love.

4

The Biblical Case for Ordaining Women

THE QUESTION OF women's ordination has become a defining issue for many people in the church today. There
are several quite different reasons for this, in the theological and cultural pressures many find urging them to go ahead
and the equal pressures many find urging them to resist this move.
There are all kinds of things one could say about these pressures,
but my task here is the more limited one of discussing some of the
key biblical texts.

In this chapter I write not about the relation between the sexes
in general, nor indeed about marriage, but about the ministry
of women. That is a welcome limitation of my subject, and I'm
going to limit it further, but I do want to set my remarks within
a particular framework of biblical theology to do with Genesis 1.
Many people have said, and I have often enough said it myself,
that the creation of man and woman in their two genders is a vital
part of what it means that humans are created in God's image. I now
regard that as a mistake. After all, not only the animal kingdom,

as noted in Genesis itself, but also the plant kingdom, as noted by the reference to seed, have their male and female. The two-gender factor is not specific to human beings but runs right through a fair amount of the rest of creation.

This doesn't mean it's unimportant; indeed, it means if anything it's all the more important. Being male and being female, and working out what that means, is something most of creation is called to do and be, and unless we are to collapse into a kind of gnosticism, where the way things are in creation is regarded as secondary and shabby compared to what we are now to do with it, we have to recognize, respect, and respond to this call of God to live in the world he has made and as the people he has made us. It's just that we can't use the argument that being male-plus-female is somehow what being God's image bearers actually means. Which brings us nicely to Galatians 3:28, and I'd like to offer some reflections on it.

Women Are Part of the Family of God

The first thing to say is fairly obvious but needs saying anyway. Galatians 3 is not about ministry. Nor is it the only word Paul says about being male and female. Instead of taking texts in a vacuum and then arranging them in a hierarchy, for instance by quoting this verse and then saying that it trumps every other verse in a kind of fight to be the senior bull in the herd (what a very masculine way of approaching exegesis, by the way!), we need to do justice to what Paul is actually saying at this point.

The point Paul is making overall in this passage is that God has one family, not two, and that this family consists of all those who believe in Jesus; that this is the family God promised to Abraham, and that nothing in the Torah can stand in the way of this unity, which is now revealed through the faithfulness of the Messiah.

This is not at all about how we relate to one another within this single family. It is about the fact, as we often say, that the ground is even at the foot of the cross.

Let me start with a note about translation and exegesis. This verse is often mistranslated such: "Neither Jew nor Greek, neither slave nor free, neither male nor female." That is precisely what Paul does *not* say; because it's what we expect he's going to say, we should note carefully what he has said instead, since he presumably means to make a point by doing so, a point that is missed when the translation is flattened out as in that version. What he says is that there is neither Jew nor Greek, neither slave nor free, *no "male and female."* I think the reason he says "no male and female" rather than "neither male nor female" is that he is actually quoting Genesis 1, and that we should understand the phrase "male and female" as a quotation.

So does Paul mean that in Christ the created order itself is undone? Is he saying, as some have suggested, that we go back to a kind of chaos in which no orders of creation apply any longer? Or is he saying that we go on, like the gnostics, from the first rather shabby creation, in which silly things like gender differentiation apply, to a new world in which we can all live as hermaphrodites—which, again, some have suggested, and which has interesting possible ethical spin-offs? No. Paul is a theologian of new creation, and it is always the renewal and reaffirmation of the existing creation, never its denial, as not only Galatians 6:15–16 but also Romans 8 and 1 Corinthians 15 make so very clear. Indeed, Genesis 1–3 remains enormously important for Paul throughout his writings.

What, then, is he saying? Remember that he is controverting in particular those who wanted to enforce Jewish regulations, and indeed Jewish ethnicity, upon Gentile converts. Remember the synagogue prayer in which the man who prays thanks God that he has not made him a Gentile, a slave, or a woman—at which point the women in the congregation thank God "that you have made me according to your will." I think Paul is deliberately marking out the family of Abraham reformed in the Messiah as a people who cannot pray that prayer, since within this family such distinctions are now irrelevant.

I think there is more. Remember that the presenting issue in Galatians is circumcision, male circumcision of course. We sometimes think of circumcision as a painful obstacle for converts, as indeed in some ways it was, but for those who embraced it, it was a matter of pride and privilege. It not only distinguished Jews from Gentiles; it marked them in a way that automatically privileged males. By contrast, imagine the thrill of equality brought about by baptism, an identical rite for Jew and Gentile, slave and free, male and female.

And that's not all. Though this is somewhat more speculative, the story of Abraham's family privileged the male line of descent: Isaac, Jacob, and so on. What we find in Paul, both in Galatians 4 and in Romans 9, is the careful attention paid—rather like Matthew 1, in fact, though from a different angle—to the women in the story. If those in Christ are the true family of Abraham, which is the point of the whole story, then the manner of this identity and unity takes a quantum leap beyond the way first-century Judaism construed it, bringing male and female together as surely and as equally as Jew and Gentile. What Paul seems to be doing in this passage, then, is ruling out any attempt to back up the continuing male privilege in the structuring and demarcating of Abraham's family by an appeal to Genesis 1, as though someone were to say, "But of course the male line is what matters, and of course male circumcision is what counts, because God made male and female." No, says Paul, none of that counts when it comes to membership in the renewed people of Abraham.

But once we have grasped this point, we must take a step back and reflect on what Paul has *not* done as well as what he has done. In regard to the Jew/Gentile distinction, Paul's fierce and uncompromising insistence on equality in Christ does not at all mean that we need pay no attention to the distinctions among those of different cultural backgrounds when it comes to living together in the church. Romans 14 and 15 are the best examples of this, but we can see it as well throughout Galatians, as Paul regularly says "we," meaning Jewish Christians, and "you" or "they" in reference to Gentile Christians. They have come to an identical destination, but they have come by very different routes and retain very different cultural memories and imaginations. The differences between them are not obliterated, and pastoral practice needs to take note of this; they are merely irrelevant when it comes to belonging to Abraham's family.

This applies, I suggest, mutatis mutandis, to Paul's treatment of men and women within the Christian family. The difference is irrelevant for membership status and membership badges. But it is still to be noted when it comes to pastoral practice. We do not become hermaphrodites or for that matter genderless, sexless beings when we are baptized. Paul would have been the first to reject the gnostic suggestion that the original creation was a secondary, poor shot at making a world and that we have to discover ways of transcending that which, according to Genesis 1, God called "very good."

This is the point at which we must issue a warning against the current fashion in some quarters, in America at least, for documents like the so-called Gospel of Mary, read both in a gnostic and a feminist light. That kind of option appears to present a shortcut right into a prowomen agenda, but it not only purchases that at a huge cost, historically and theologically, but also presents a two-edged blessing, granted the propensity in some branches of ancient gnosticism to flatten out the male/female distinction not by affirming both as equally important but by effectively turning women into men. The last saying in the so-called Gospel of Thomas suggests that "Mary will be saved if she makes herself male." That presents a radically different agenda from what we find in the New Testament.

Women Leaders in the Early Church

Among the many things that need to be said about the Gospels is that we gain nothing by ignoring the fact that Jesus chose twelve male apostles. There were no doubt all kinds of reasons for this within both the symbolic world in which he was operating and the practical and cultural world within which they would have to live and work. But every time this point is made—and in my experience it is made quite frequently—we have to comment on how interesting it is that there comes a time in the story when the disciples all forsake Jesus and run away; at that point, long before the rehabilitation of Peter and the others, it is the women who come first to the tomb, who are the first to see the risen Jesus, and who are the first to be entrusted with the news that he has been raised from the dead.

This is of incalculable significance. Mary Magdalene and the others are the apostles to the apostles. We should not be surprised that Paul calls a woman named Junia an apostle in Romans 16:7. If an apostle is a witness to the resurrection, there were women who deserved that title before any of the men. (I note that there was a huge fuss in the translation and revision of the New International Version at the suggestion that Junia was a woman and not a single historical or exegetical argument was available to those who kept insisting, for obvious reasons, that she was Junias, a man.)

Nor is this promotion of women totally new with the resurrection. As in so many other ways, what happened then picked up hints and pinpoints from earlier in Jesus's public career. I think in particular of the woman who anointed Jesus (without here going into the question of who it was and whether it happened more than once); as some have pointed out, this was a priestly action that Jesus accepted as such.

I think too of the remarkable story of Mary and Martha in Luke 10. Most of us grew up with the line that Martha was the

active type and Mary the passive or contemplative type, and that
Jesus is simply affirming the importance of both and even the
priority of devotion to him. That devotion is undoubtedly part
of the importance of the story, but far more obvious to any first-
century reader, and to many readers in Turkey, the Middle East,
and many other parts of the world to this day, would be the fact
that Mary was sitting at Jesus's feet *in the male part of the house*
rather than being kept in the back rooms with the other women.
This, I am pretty sure, is what really bothered Martha; no doubt
she was cross at being left to do all the work, but the real problem behind that was that Mary had cut clean across one of the
most basic social conventions. It is as though, in today's world,
you were to invite me to stay in your house and, when it came to
bedtime, I were to put up a camp bed in your bedroom. We have
our own clear but unstated rules about whose space is which. So
did they, and Mary has just flouted them. *And Jesus declares that she
is right to do so.* She "sat at the master's feet," a phrase that doesn't
mean what it would mean today—the adoring student gazing up
in admiration and love at the wonderful teacher. As is clear from
the use of the phrase elsewhere in the New Testament (for instance, Paul with Gamaliel), to sit at the teacher's feet is a way of
saying you are being a *student* and picking up the teacher's wisdom
and learning; in that very practical world, you wouldn't do this
just for the sake of informing your own mind and heart, but in
order to become yourself a teacher, a rabbi.

Like much in the Gospels, this story is left cryptic as far as we at
least are concerned, but I doubt if any first-century reader would
have missed the point. That, no doubt, is at least part of the reason we find so many women in positions of leadership, initiative,
and responsibility in the early church. I used to think Romans 16
was the most boring chapter in the letter, and now, as I study the
names and think about them, I am struck by how powerfully they

indicate the way the teaching both of Jesus and of Paul was being worked out in practice.

One other point, about Acts, an insight among many others that I gleaned from Ken Bailey on the basis of his long experience of working in the Middle East. It's interesting that at the crucifixion the women were able to come and go and see what was happening without fear of the authorities. They were not regarded as a threat and did not expect to be so regarded. Bailey points out that this pattern is repeated to this day in the Middle East; at the height of the troubles in Lebanon, when men on all sides in the factional fighting were either hiding or going about with great caution, women were free to come and go, to do the shopping, to take children out, and so on. It's fascinating, then, that when we turn to Acts and the persecution that arose against the church not least at the time of Stephen, we find that women are being targeted equally alongside the men. Saul of Tarsus was going to Damascus to catch women and men alike and haul them off into prison. Bailey points out on the basis of his cultural parallels that this only makes sense if the women too are seen as leaders, influential figures within the community.

Decoding the Challenging Passages in 1 Corinthians

An enormous amount of work has been done recently on the social and cultural context of 1 Corinthians, and I want to urge all those who are interested in finding out what Paul actually said and meant to study such work with great care. There are many things about first-century classical life that shed a great deal of light on the actual issues Paul is addressing, and they need to be taken carefully into account.

"The Women Should Keep Silence"

I want to home in at once on one of the two passages that have caused so much difficulty, the verses at the end of 1 Corinthians 14 in which Paul insists that women must keep silent in church. I am of two minds whether to agree with those who say this verse is a later and non-Pauline interpolation. One of the finest textual critics of our day, Gordon Fee, has argued strongly that it is, purely on the grounds of the way the manuscript tradition unfolds. I urge you to examine his arguments and make up your own mind.

But I have always been attracted, ever since I heard it, to the explanation offered by Ken Bailey. In the Middle East, he says, it was taken for granted that men and women would sit apart in church, as still happens today in some circles. Equally important, the service would be held (in Lebanon, say, or Syria, or Egypt) in formal or classical Arabic, which the men would all know but which many of the women would not, since the women would speak only a local dialect or patois. Again, we may disapprove of such an arrangement, but one of the things you learn in real pastoral work as opposed to ivory-tower academic theorizing is that you simply can't take a community all the way from where it currently is to where you would ideally like it to be in a single flying leap.

Anyway, the result would be that during the sermon in particular, the women, not understanding what was going on, would begin to get bored and talk among themselves. As Bailey describes the scene in such a church, the level of talking from the women's side would steadily rise in volume, until the minister would have to say loudly, "Will the women please be quiet!" whereupon the talking would die down but only for a few minutes. Then, at some point, the minister would again have to ask the women to be quiet, and he would often add that if they wanted to know what was being said, they should ask their husbands to explain it to them when they got home. I know other explanations are sometimes offered for this passage, some of them quite plausible; this is the one that has struck me for many years as having the strongest claim to provide a context for understanding what Paul is saying. After all, his central concern in 1 Corinthians 14 is for order and decency in the church's worship. This would fit extremely well.

What the passage cannot possibly mean is that women had no part in leading public worship, speaking out loud as they did so. This positive point is proved at once by the other relevant Corinthian passage, 1 Corinthians 11:2–11, since there Paul gives instructions on how women are to dress while engaging in such activities, instructions that obviously wouldn't be necessary if they had been silent in church all the time. But that is the one thing we can be sure of. In this passage, almost everything else seems to me remarkably difficult to nail down. What I want to do now is to offer you the explanation I tried out in *Paul for Everyone: 1 Corinthians*. There is more to be said, no doubt, but probably not less.

Paul's Directive Regarding Head Coverings

Paul wasn't, of course, addressing the social issues we know in our world. Visit a different culture, even today, and you will discover many subtle assumptions, pressures, and constraints in society, some of which appear in the way people dress and wear their hair. In Western culture, a man wouldn't go to a dinner party wearing a bathing suit, nor would a woman attend a beach picnic wearing a wedding dress. Most Western churches have stopped putting pressure on women to wear hats in church (Western-style hats, in any case, were not what Paul was writing about here), but nobody thinks it odd that we are still strict about men *not* wearing hats in church.

In Paul's day (as, in many ways, in ours), gender was marked by hair and clothing styles. We can tell from statues, vase paintings, and other artwork of the period how this worked out in practice. There was social pressure to maintain appropriate distinctions. But did not Paul himself teach that there was "no 'male and female'; you are all one in the Messiah" (Galatians 3:28)? Perhaps, indeed, that was one of the traditions that he had taught the Corinthian church, where churchgoers needed to know that Jew and Greek, slave and free, male and female were all equally welcome, equally valued, in the renewed people of God. Perhaps that had actually created the situation he addresses here; perhaps some of the Corinthian women had been taking him literally, so that when they prayed or prophesied aloud in church meetings (which Paul assumes they would do regularly; this tells us, as we've seen, something about how to understand 1 Corinthians 14:34–35), they had decided to remove their normal head covering, perhaps also unbraiding their hair, to show that in the Messiah they were free from the normal social conventions by which men and women were distinguished.

That's a lot of *perhaps*es. We can only guess at the dynamics of the situation—which is what historians always do. It's just that here we are feeling our way in the dark more than usual. Perhaps to the Corinthians' surprise, Paul doesn't congratulate the women on this new expression of freedom. He insists on maintaining gender differentiation during worship.

Another dimension to the problem may well be that in the Corinth of his day the only women who appeared in public without some kind of head covering were prostitutes. This isn't suggested directly here, but it may have been in the back of his mind. If the watching world discovered that the Christians were having meetings where women "let their hair down" in this fashion, it could have the same effect on their reputation as it would in the modern West if someone looked into a church and found the women all wearing bikinis.

The trouble is, of course, that Paul doesn't say exactly this, and we run the risk of explaining him in terms that (perhaps) make sense to us while ignoring what he himself says. It's tempting to do that, precisely because in today's Western world we don't like the implications of the differentiation he maintains in 1 Corinthians 3:11 the Messiah is the head of every man, a husband is the head of every woman, and the head of the Messiah is God. This seems to place man in a position of exactly that assumed superiority against which women have rebelled, often using Galatians 3:28 as their battle cry.

But what does Paul mean by *head*? He uses it here sometimes in a metaphorical sense, as in 1 Corinthians 3:11, and sometimes literally, as when he's talking about what to do with actual human heads (verses 4–7 and 10). But the word can mean various things, and a good case can be made that in verse 3 he is referring not to headship in the sense of sovereignty but to headship in the sense of "source," like the source or head of a river. In fact, in some of the key passages where he explains what he's saying (verses 8, 9, and 12a), he refers explicitly to the creation story in Genesis 2, where woman was made from the side of man. I suspect, in fact, that this is quite a different use of the idea of headship from that in Ephesians 5, where it relates to husband and wife and a different point is being made. That doesn't mean Paul couldn't have written them both, only that he was freer than we sometimes imagine to modify his metaphors according to context.

The underlying point seems to be that in worship it is important for both men and women to be their truly created selves, to honor God by being what they are and not blurring the lines by pretending to be something else. One of the unspoken clues to this passage may be Paul's assumption that in worship the creation is being restored, or perhaps that in worship we are anticipating its eventual restoration (15:27–28). God made humans male and female, and gave them authority over the world, as Ben-Sirach 17:3–4 puts it, summarizing Genesis 1:26–28 and echoing Psalm 8:4–8 (Ben-Sirach was written around 200 BC). And if humans are to reclaim this authority over the world, this will come about as they worship the true God, as they pray and prophesy in his name and are renewed in his image, in being what they were made to be, in celebrating the genders God has given them.

If this is Paul's meaning, the critical move he makes is to argue that a man dishonors his head by covering it in worship and that a woman dishonors hers by *not* covering it. He argues this mainly on the basis that creation itself tends to give men shorter hair and women longer (1 Corinthians 11:5–6, 13–15); the fact that some cultures, and some people, offer apparent exceptions would probably not have worried him. His main point is that in worship men should follow the dress and hair codes that proclaim them to be male, and women the codes that proclaim them to be female.

Why then does he say that a woman "must have authority on her head because of the angels" (verse 10)? This is one of the most puzzling verses in a puzzling passage, but there is help of sorts in the Dead Sea Scrolls. There is it assumed that when God's people meet for worship, the angels are there too (as many liturgies and theologians still affirm). For the scrolls, this means that the angels, being holy, must not be offended by any appearance of unholiness among the congregation. Paul shares the assumption that angels worship along with humans but may be making a different point.

When humans are renewed in the Messiah and raised from the dead, they will be set in authority over the angels (6:3). In worship, the church anticipates how things are going to be in that new day. When a woman prays or prophesies (perhaps in the language of angels, as in 13:1), she needs to be truly what she is, since it is to male and female alike, in their mutual interdependence as God's image-bearing creatures, that the world, including the angels, is to be subject. God's creation needs humans to be fully, gloriously, and truly human, which means fully and truly male and female. This and of course much else besides is to be glimpsed in worship.

The Corinthians, then, may have drawn the wrong conclusion from the tradition that Paul had taught them. Whether or not they followed his argument any better than we can, it seems clear that his main aim was that marks of difference between the sexes should not be set aside in worship. This is the best sense I can see in this admittedly difficult passage.

We face different issues, but making sure our worship is ordered appropriately, to honor God's creation and anticipate its fulfillment in the new creation, is still a priority. There is no "perhaps" about that. When we apply this to the question of women's ministry, it seems to me that we should certainly stress equality in the role of women but should be very careful about implying identity. This passage falls, for me at least, quite strongly on the side of those who see the ministry of women as significantly different from the ministry of men and therefore insists that we need both sexes to be themselves, rather than for one to try to become a clone of the other.

All this points us toward the final and hardest passage of all, 1 Timothy 2.

Decoding the Challenging Passages in 1 Timothy

So this is what I want: the men should pray in every place, lifting up holy hands, with no anger or disputing. In the same way the women, too, should clothe themselves decently, being modest and sensible about it. They should not go in for elaborate hairstyles, or gold, or pearls, or expensive clothes. Instead, as is appropriate for women who profess to be godly, they should adorn themselves with good works. They must study undisturbed, in full submission to God. I'm not saying that women should teach men, or try to dictate to them; rather, that they should be left undisturbed. Adam was created first, you see, and then Eve; and Adam was not deceived, but the woman was deceived, and fell into trespass. She will, however, be kept safe through the process of childbirth, if she continues in faith, love, and holiness with prudence. (1 Timothy 2:8–15)

I leave completely aside for today the question of who wrote 1 Timothy. It diverges more sharply from the rest of Paul than any of the other letters, including the other pastorals and 2 Thessalonians. But I do not discount it for that reason; many of us write in different styles according to the occasion and audience, and though that doesn't remove all the problems, it ought to contextualize them. What matters, and matters vitally in a great many debates, is of course what the passage says. I don't think I exaggerate when I suggest that this passage above all others has been the sheet anchor for those who want to deny women a place in the ordained ministry of the church, with full responsibilities for preaching, presiding at the Eucharist, and exercising leadership within congregations and indeed dioceses.

Once again the matter is vexed and much fought over, and I have not read more than a fraction of the enormous literature that has been produced on the passage. I simply give my opinion for what it is worth. And once again I draw here on what I have said in my recent popular-level commentary on the passage (*Paul for Everyone: The Pastoral Epistles*). That commentary goes with and explains my translation of the passage, which draws out some ways in which the words can actually mean something significantly different from what has usually been assumed.

Women Teaching Men

When people say that the Bible enshrines patriarchal ideas and attitudes, this passage, particularly verse 12, is often held up as the prime example. Women mustn't be teachers, the verse seems to say; they mustn't hold authority over men; they must keep silent. That, at least, is how many translations put it. This is the main passage that people quote when they want to suggest that the New Testament forbids the ordination of women. I was once reading these verses in a church service when a woman near the front exploded in anger, to the consternation of the rest of the congregation (even though some agreed with her). The whole passage seems to say that women are second-class citizens at every level. They aren't even allowed to dress attractively. They are the daughters of Eve, and she was the original troublemaker. The best thing for them to do is to get on and have children, to behave themselves and keep quiet.

Well, that's how most people read the passage in our culture until quite recently. I fully acknowledge that the very different reading I'm going to suggest may sound initially as though I'm simply trying to make things easier, to tailor this bit of Paul to fit our culture. But there is good, solid scholarship behind what I say, and I genuinely believe it may be the right interpretation.

When you look at cartoon strips, B-grade movies, and Z-grade novels and poems, you pick up a standard view of how everyone imagines men and women behave. Men are macho, loud-mouthed, arrogant thugs, always fighting and wanting their own way. Women are simpering, empty-headed creatures, who think about nothing except clothes and jewelry. There are Christian versions of this, too: men must make the decisions, run the show, always be in the lead, telling everyone what to do; women must stay at home and bring up the children. If you start looking for a biblical backup for this view, well, what about Genesis 3? Adam would never have sinned if Eve hadn't given in first. Eve has her punishment, and it's pain in childbearing (Genesis 3:16).

You don't have to embrace every aspect of the women's liberation movement to find that interpretation hard to swallow. Not only does it stick in our throats as a way of treating half the human race; it doesn't fit with what we see in the rest of the New Testament, in the passages we've already glanced at.

The key to the present passage, then, is to recognize that it commands that women, too, should be allowed to study and learn, and should not be restrained from doing so (verse 11). They are to be "in full submission"; this is often taken to mean "to the men" or "to their husbands," but it is equally likely that it refers to the learner's attitude of submission to God or to the gospel—which of course would also be true for men. Then the crucial verse 12 need not be read as "I do not allow a woman to teach or hold authority over a man"—the translation that has caused so much difficulty in recent years. It can equally mean (and in context this makes much more sense): "I don't mean to imply that I'm now setting up women as the new authority over men in the same way that previously men held authority over women." Why might Paul need to say this?

There are some signs in the letter that it was originally sent to Timothy while he was in Ephesus. And one of the main things we know about religion in Ephesus is that the main religion—the biggest temple, the most famous shrine—was a female-only cult. The Temple of Artemis (that's her Greek name; the Romans called her Diana) was a massive structure that dominated the area. As befitted worshippers of a female deity, the priests were all women. They ruled the show and kept the men in their place.

Now, if you were writing a letter to someone in a small, new religious movement with a base in Ephesus, and you wanted to say that because of the gospel of Jesus the old ways of organizing male and female roles had to be rethought from top to bottom, with one feature being that women were to be encouraged to study and learn and take a leadership role, you might well want to avoid giving the wrong impression. Was the apostle saying, people might wonder, that women should be trained so that Christianity would gradually become a cult like that of Artemis, where women led and kept the men in line? That, it seems to me, is what verse 12 is denying. The word I've translated as "try to dictate to them" is unusual but has overtones of "being bossy" or "seizing control." Paul is saying, like Jesus in Luke 10, that women must have the space and leisure to study and learn in their own way, not in order that they may muscle in and take over the leadership as in the Artemis cult, but so that men and women alike can develop whatever gifts of learning, teaching, and leadership God has given them.

What's the point of the other bits of the passage, then? Verse 8 is clear: the men must give themselves to devout prayer and must not follow the normal stereotypes of male behavior: no anger or arguing. Then verses 9 and 10 make the same point about women: they must be set free from their stereotype, that of fussing all the time about hairdos, jewelry, and fancy clothes—but not to become dowdy, unobtrusive little mice but so that they can make a creative contribution to the wider society. The phrase "good works" in verse 10 sounds bland to us, but it's one of the regular ways people used to refer to the social obligation to spend time and money on people less fortunate than oneself, to be a benefactor of the town through helping public works, the arts, and so on.

Why does Paul finish off with the explanation about Adam and Eve? Remember that his basic point is to insist that women, too, must be allowed to learn and study as Christians and not be kept in unlettered, uneducated boredom and drudgery. The story of Adam and Eve makes the point well: look what happened when Eve was deceived. Women need to learn just as much as men do. Adam, after all, sinned quite deliberately; he knew what he was doing and that it was wrong, and he went ahead. The Old Testament is stern about that kind of action.

And what of the bit about childbirth? Paul doesn't see it as a punishment. Rather, he offers assurance that, though childbirth is indeed difficult, painful, and dangerous, often the most testing moment in a woman's life, this is not a curse to be taken as a sign of God's displeasure. God's salvation is promised to all, women and men, who follow Jesus in faith, love, holiness, and prudence. And that salvation is promised to those who contribute to God's

creation through childbearing, just as it is to everyone else. Becoming a mother is hard enough, God knows, without pretending
it's somehow an evil thing. Let's not leave any more unexploded
bombs and mines for people to blow their minds with. Let's read
this text as I believe it was intended, as a way of building up
God's church, men and women, women and men alike. And just
as Paul was concerned to apply this in one particular situation, so
we must think and pray carefully about where our own cultures,
prejudices, and angers are taking us, and make sure we conform
not to the stereotypes the world offers but to the healing, liberating, humanizing message of the gospel of Jesus.

IT IS HIGH TIME to sum up. I think I have said enough to show
you where I think the evidence points. I believe we have seriously
misread the relevant passages in the New Testament, not least
through a long process of assumption, tradition, and all kinds of
postbiblical and subbiblical attitudes that have crept into Christianity. Just as I think we need to radically change our traditional
pictures of the afterlife, away from medieval models and back to
biblical ones, so we need to radically change our traditional pictures both of what men and women are and of how they relate
to one another within the church, and indeed of what the Bible
says on this subject. I do wonder, sometimes, if those who present radical challenges to Christianity have not been all the more
eager to make out that the Bible says certain things about women
as an excuse for claiming that Christianity in general is a wicked
thing that should be abandoned. Of course, plenty of Christians
have given outsiders enough chances to make that sort of comment. But perhaps in our generation we have an opportunity to
take a large step back in the right direction.

5

Jesus Is Coming—
Plant a Tree!

MY INTENTION IN this chapter is, as always, to pay the fullest possible attention to scripture in both its details and its broad sweep, and to allow the biblical writers to set the agenda rather than forcing on them a scheme of thought that does not do them justice. This task is made harder still by the traditions of thought, prayer, spirituality, and ethics in various parts of the church. Here again my aim is always to allow scripture to enter into dialogue with traditions, including those traditions that think of themselves as biblical, and to critique them when they are less than fully in accordance with scripture. It is my belief that the broad sweep of Western theology since way before the Reformation, and continuing since the sixteenth century in both Roman Catholicism and the various branches of Protestantism, has been subbiblical in its approach to that potent combination of themes, eschatology, and ecology.

We have declared, in the Nicene Creed, that Jesus Christ "will come again in glory to judge the living and the dead, and his kingdom shall have no end," but neither mainline Catholic nor

mainline Protestant theology has explored what exactly we mean
by all that, and we have left a vacuum to be filled by various kinds
of dualism. In particular, Western Christianity has allowed itself
to embrace that dualism whereby the ultimate destiny of God's
people is heaven, seen as a place detached from earth, so that the
aim of Christianity as a whole, and of conversion, justification,
sanctification, and salvation, is seen in terms of leaving earth behind and going home to a place called heaven.

So powerful is this theme in a great deal of popular preaching, liturgy, and hymnography that it comes as a shock to many
people to be told that this is simply not how the earliest Christians
saw things. For the early Christians, the resurrection of Jesus
launched God's new creation upon the world, beginning to fulfill
the prayer Jesus taught his followers, that God's kingdom would
come "on earth as in heaven" (Matthew 6:10), and anticipating
the "new heavens and a new earth" (Isaiah 65:17; 66:22; 2 Peter
3:13; Revelation 21:1) promised by Isaiah and again in the New
Testament. From this point of view, as I have often said (though
the phrase is not original to me), heaven is undoubtedly important, but it's not the end of the world. The early Christians were
not very interested, in the way our world has been interested, in
what happens to people immediately after they die. They were
extremely interested in a topic many Western Christians in the
last few hundred years have forgotten about altogether, namely the
final new creation, new heavens and new earth joined together,
and the resurrection of the body that will create new human beings to live in that new world.

The question of how you think about the ultimate future has
an obvious direct impact on how you think about the task of the
church in the present time. To put it crudely and at the risk of
caricaturing: if you suppose that the present world of space, time,
and matter is a thoroughly bad thing, then the task is to escape
from this world and enable as many others to do so as possible. If
you go that route, you will most likely end up in some form of

gnosticism, and the gnostic has no interest in improving the lot of human beings, or the state of the physical universe, in the present time. Why wallpaper the house if it's going to be knocked down tomorrow?

At the opposite end of the spectrum, some theologians have been so impressed with the presence and activity of God in the present world that they have supposed God wants simply to go on working at it as it is, to go on improving it until eventually it becomes the perfect place he has in mind. From this point of view, the task of the Christian is to work at programs of social and cultural improvement, including care for the natural environment, so that God's kingdom will come on earth through an almost evolutionary process, as in Teilhard de Chardin, or at least until human hard work in the present world attains the result God ultimately intends.

As I said, these are caricatures, but they do exist in reality here and there, and those who incline toward one or the other regularly sustain their end of the spectrum by pointing out the folly of those who live at the other end. It won't surprise you to know that I take a position at neither extreme. I don't just want to occupy a kind of middle ground, but rather the high ground of a biblical eschatology, which transcends altogether the kind of either/or such a spectrum might indicate.

But before I get to business, let me illustrate just a little. I first ran into the problem I'm addressing here during a weekend of lectures in Thunder Bay, Ontario, in (I think) 1982 or 1983. I was working in Montreal at the time and was asked to talk about Jesus in historical context, a subject I'd been lecturing about that eventually turned into my book *Jesus and the Victory of God*. But to my surprise, the main question people had in mind was not the meaning of the parables or of the cross or the incarnation itself, but questions of ecology: some people in the church had been saying that there was no point in worrying about the trees and acid rain, the rivers and lakes and water pollution, or climate change in relation to crops and harvests, because Jesus was coming back soon and Armageddon would destroy the present world. Not only was there no point in being concerned about the state of the ecosystem; it was actually unspiritual to do so, a form of worldliness that distracted from the real task of the gospel, which was the saving and nurturing of souls for a spiritual eternity. I can't now remember what sort of answers I gave to these questions, but the questions themselves have stayed with me.

Where then do we go to articulate the rich and deep New Testament eschatology that will help us make sense of Christian responsibility within the present world? One place above all, a chapter that towers above most others and without which I personally would be quite lost within my worldview: Romans 8.

Romans 8

Romans 8 is so rich and dense, so full of comfort and challenge, that it's easy to get lost and to fail both to see the forest for the trees and, in some cases, to see some particular trees because of the forest. I begin with the key passage, verses 18–27, then I pull the camera back a bit, widen the angle as it were, so we can see the larger narrative within which this forms the great climax.

> This is how I work it out. The sufferings we go through in the present time are not worth putting in the scale alongside the glory that is going to be unveiled for us. Yes: creation itself is on tiptoe with expectation, eagerly awaiting the moment when God's children will be revealed. Creation, you see, was subjected to pointless futility, not of its own volition, but because of the one who placed it in this subjection, in the hope that creation itself would be freed from its slavery to decay, to enjoy the freedom that comes when God's children are glorified. Let me explain.

We know that the entire creation is groaning together,
and going through labor pains together, up until the present time. Not only so: we too, we who have the first fruits
of the spirit's life within us, are groaning within ourselves, as we eagerly await our adoption, the redemption
of our body. We were saved, you see, in hope. But hope
isn't hope if you can see it! Who hopes for what they can
see? But if we hope for what we don't see, we wait for it
eagerly—but also patiently. In the same way, too, the spirit
comes alongside and helps us in our weakness. We don't
know what to pray for as we ought to; but that same spirit
pleads on our behalf, with groanings too deep for words.
And the Searcher of Hearts knows what the spirit is thinking, because the spirit pleads for God's people according to
God's will. (Romans 8:18–27)

I said that this is the climax of the letter so far, and that may
itself be controversial. As a lifelong student of Romans and of
commentaries on it, I am impressed by—and depressed by!—the
strength of the tradition that has effectively marginalized these
verses. On the face of it, purely in terms of the flow of Paul's
argument, verses 18–27 stand near what must be counted the rhe-
torical climax of this, the central section of Paul's longest and
most intricately argued letter. And yet, as I say, preachers, commentators, and theologians in the Western tradition, both Catholic and Protestant, have almost routinely regarded this section as
something of a distraction. We know, so it seems, that Romans is
basically about how individuals get saved, that is, find their way to
heaven at last, having been justified by faith and led by the Spirit
into a life of holiness. So our eyes skip over the difficult language
in verses 19–23, the heart of the passage, and go on from the call
of holiness in verses 13–17 to the assurance of salvation in verses
28–30, relishing the matchless coda in verses 31–39 and forgetting
the awkward bit in the middle. But, as usual when we do this

kind of thing with Paul, we lose something absolutely vital, in this case something that calls into question an entire worldview.

God's Good Creation

Paul is talking about the glory that, he says, is to be unveiled "for us" (verse 18). What he means by that is instantly explained, as is his wont, in the next verse, which is then elaborated in the next two. The whole creation, the entire cosmos, is on tiptoe with expectation for God's children to be revealed. Glory is not simply a kind of luminescence, as though the point of salvation were that we would eventually shine like electric lightbulbs. *Glory* means, among other things, rule and power and authority; as other writers (notably Saint John the Divine) make clear, part of the point of God's saving his people is that they are destined not merely to enjoy a relaxing endless vacation in a place called heaven, but that they are designed to be God's stewards, ruling over the whole creation with healing and restorative justice and love. Verses 20 and 21 make this plain. The whole creation has become, by God's decree, a place of corruption in both senses—deterioration both moral and physical. We often think of the universe of space, time, and matter as automatically corruptible, but Paul insists this was not the creator's original intention. As always when he's dealing with the ultimate future, he goes back to Genesis 1, 2, and 3 so that his eschatology is rooted in a strong view of creation as God's good handiwork.

Then comes the great statement of the ultimate future in verse 21: the creation itself will be set free from the slavery that consists in corruption. That freedom will come when God's children are glorified. Many translations obscure this, implying a generalized freedom and glory to be shared equally by the redeemed and by creation, but this is not what Paul says and certainly not what he means. What he means, as in verses 18 and 19, is that when God's redeemed people are finally rescued, which as we will presently discover means when we are given our resurrection bodies, we will be set in "glory," that is, sovereign rule as God's image-bearing children (compare verse 29), over the whole creation—and then, at last, God's project will be where it was supposed to be going right at the start, when God created humankind in his own image, to be fruitful and to look after the garden. When the humans are put right, creation will be put right. That is the ultimate point, the glorious full sweep of Paul's soteriology.

A brief reflection at this point in dialogue with the Western tradition. Paul criticizes his fellow Jews for failing to see that the reason God called them in the first place was so that through them he could rescue the whole human race. They were right to regard their call as nonnegotiable but wrong to think it was only about them. The same thing seems to have happened at the next stage, in mainline Christian readings of Paul. We have been right, deeply right, to think that Paul is vitally concerned with the salvation of human beings and all that goes with that: the redeeming death of Jesus, justification by grace through faith, and so on. But we have been wrong to suppose that the only purpose was the salvation of humans—as it were, away from the world, away from the whole created order. As Israel was in God's purposes vis-à-vis the whole human race, so are humans vis-à-vis the whole created order. That's how it has been since Genesis 1 and 2, and that's how, according to Paul, God intends it to be at the last.

And that is why, in the passage that follows, Paul describes the concentric circles of "groaning": the world groaning in travail, like a woman in labor awaiting the birth of the child; the church groaning with inexpressible sighs, longing for the resurrection of the body; and—far and away the most remarkable thing of all—the Spirit groaning within us in ways too deep for articulate speech. I have written elsewhere that this is no incidental reference to prayer and the work of the Spirit. The whole point is that when we pray we are not merely distant or feeble petitioners. We are starting to take up our responsibility as God's image-bearing human beings, sharing God's rule over creation. Because we ourselves are weak and corrupted in our unresurrected state, the exercise of this glory is always mysterious and even painful, as we find ourselves in prayer at the heart of the pain of the world, and yet discover God's spirit groaning within us.

That is a key part of our calling in the present time and very relevant to our topic. We should take responsibility in the present time for God's groaning creation. This responsibility will be focused particularly in prayer, even if we can't put our deepest concerns into articulate speech. This responsibility rests upon an analysis of God's world, of the whole creation (nature, as we say, "red in tooth and claw"). We are not to regard the created order as random, nor to see its present disarray as somehow its own fault, but to understand it in terms of an initially good creation now radically spoiled but awaiting redemption. There is nothing wrong with space, time, and matter; what is wrong is what's happened to this good creation. It is enslaved, and what slaves need is redemption.

Redemption Because of God's Righteousness

The theme of slaves being redeemed provides the clue that will help us grasp the larger picture that Paul has been developing for the last few chapters, the picture that ought to frame our conversations about a Christian view of the whole created order in the future and in the present. Like most Jews throughout history, Paul was far more aware than we are ever likely to be of the story not only of the creation but also of the Exodus, the time when the creator God went down to Egypt to rescue his people from slavery and lead them home through the wilderness to their promised land, guiding them to that inheritance by his own personal presence. Among the key stages of this divine work are the crossing of the Red Sea, as the people were brought through the water out of slavery and into freedom, and the arrival at Mount Sinai for the giving of the Law. All this happened as the long fulfillment of the promises God made to Abraham in Genesis 15, where he established a covenant with him and his family, requiring Abraham to believe that God was indeed the creator and life giver, the one who gives life to the dead and calls into existence things that are nonexistent.

This aspect of God's character, the promise-keeping covenant faithfulness by which he will put his world to rights, can be summed up in a single phrase: God's righteousness. And with that, working back step-by-step from the language of Romans 8:18–27, we realize that we have been describing the whole theme of the letter to the Romans. Romans is not just about me, my sin, and my salvation, though to be sure it certainly is about that too, and centrally. It is about the way the creator God keeps the covenant he made with Abraham by providing an Exodus to beat all Exoduses.

Now watch how this works out in the detail of Romans— and realize what a powerful emphasis this places on the God-given human responsibility for creation. Having announced and explored his main theme of God's righteousness—that's a topic for another time!—in chapters 1–3, Paul grounds it in Romans 4 by expounding the promises of God to Abraham and the faith by which Abraham grasped those promises. Summarizing these promises, Paul declares—all the more remarkably considering that the promises were normally in terms of the Holy Land, the land of Israel—that God's promise to Abraham was that he would inherit the *world*. Paul, in company with some other Jewish thinkers of the time, looks beyond the temporary and limited reach of a territorially restricted promise to God's ultimate intention (as already, for instance, in Psalm 72 and parts of Isaiah). The whole world, he is declaring, is now God's holy land.

This then sets the scene for the story of the great, ultimate

Exodus—the Exodus of the human race and, with it, all creation. Paul begins with Adam in Romans 5, in slavery to sin with his descendants. Then, in chapter 6, he describes how they come through the water from slavery to freedom, through baptism into Christ's death and resurrection. This brings them, naturally, to Mount Sinai and the giving of the Law, only now it becomes clear that the Law, although it's God's holy and righteous law, must condemn them. Like Israel in the wilderness, they are destined to perish because of their disobedience, were it not for the fresh act of grace by which God rescues them: what the Law could not do, because it was weak through the flesh, God has done in the death of Jesus Christ and the power and presence of the Spirit.

Romans 8 thus comes up, as it were, in three dimensions. Paul's emphasis on the indwelling of the Spirit in 8:1–11 corresponds directly to the way in which, as the children of Israel were on their way through the wilderness, God dwelt in their midst in the pillar of cloud and fire; the result, as in Jewish theology, is the rebuilding of the temple or tabernacle, which in the terms of Romans 8 means nothing more nor less than the resurrection of the body (8:9–11). There is much more to say about all this, but not here and now.

The Promised Inheritance and Our Present Responsibility

This brings us to the crucial passage, Romans 8:12–17. Here Paul speaks of Christians being "led by the Spirit" and so refusing the opportunity to go back again into slavery and fear (8:14–15)—just like Israel in the wilderness, fantasizing about going back to Egypt and being told instead to go on, led by the presence of God in their midst. That presence assures them, again just like Israel, that they are God's children: "Israel is my firstborn son," the great slogan of the Exodus, is answered here by "You received the spirit of sonship, in when we call out, 'Abba, Father'" (8:15).

Thus it is that the Spirit bears witness with our spirit that we are God's "children. And if we're children, we are also heirs of God" (8:16–17), those who will receive the promised inheritance (compare Galatians 4:1–7; Ephesians 1:14).

But what is the inheritance? Here centuries of the Western Christian tradition have given the emphatic, though often implicit, reply: heaven. Heaven is our home, our inheritance; we have reread the story of the Exodus in those terms, with the crossing of the Jordan symbolizing, as in *Pilgrim's Progress,* by John Bunyan, the bodily death that will bring us to heaven itself, the Canaan for which we long. But—and this, as you will realize, is the whole point I've taken so long to get to—*this is precisely not what Paul says.* What he says would have been clear had not the whole Western tradition been determined to look the other way at the crucial point. The inheritance is not heaven. Nor is it Palestine, a small geographical strip in the Middle East. The inheritance is the whole renewed, restored creation. I will say it again: *the whole world is now God's holy land.* That is how Paul's retold Exodus narrative makes full and complete sense. And that, I suggest, is the ground plan for a fully biblical, fully Christian view of creation and of our responsibility toward it.

Let me just spell this out before turning to my second major topic, that of the Second Coming. Even after the reading of Romans I have offered, someone might still say, "Well, maybe one day God will renew the whole created order; that will be marvelous, and I look forward to sharing in it with my resurrection body. But I don't have that resurrection body yet, so what can I hope to do to help creation in its present state of groaning? And isn't it a bit presumptuous to think there's anything I can do, if in fact God will eventually do it all himself? Shouldn't we just wait until then?" Well, there might be some logic to that. But the whole point of the Christian gospel, as I will emphasize later, is that with the resurrection of Jesus and the gift of the Spirit, God's future has come forward to meet us in the present; what God intends to do at the last has already broken into the world the way it is. (That, indeed, is the really dramatic point about Easter: not that God has performed an extraordinary miracle on Jesus's behalf, but that with Jesus the new creation has already begun, and Jesus's followers are invited not only to benefit from it but to share in the new project it unleashes.) We are not in the position of first-century Jews, waiting for the great eschatological event; we are with the early Christians, celebrating the fact that part one of that great event has already happened in the resurrection of Jesus and the outpouring of the Spirit, and looking forward eagerly to part two, when what began at Easter and Pentecost will at last be completed.

Thus with Easter and Pentecost the remaking of God's creation of space, time, and matter has already begun. One of the primary places where this remaking is to be seen and known is in Christian holiness, which is not a matter of observing rules and regulations but of Christians taking proper human responsibility for a bit of the created order, their own bodies, and working at making them reflect the image of God as they were made to (see, again, 8:29). But precisely because bearing God's image doesn't just mean sorting out yourself but also reflecting that image into the creation in wise stewardship, the renewal of human life to which we are called in Romans 8:12–17 must issue and does issue in Christian responsibility for creation, in the present as well as the future.

Look at it like this. If someone came to you and said he or she was having real trouble resisting temptation and was always falling helplessly into sin, but that it didn't matter because one day God would provide a new body that wouldn't be capable of sinning, so why not wait for that, I hope that you would respond with a sharp dose of inaugurated eschatology. God, you would say, has *already* begun that ultimate, final work of new creation; by baptism and faith you have left behind the old order of sin and death, and by God's spirit within you, you have God's own resurrection power to enable you, even in the present, to resist sin and live as a fully human being at last; you must therefore live, in the present, as far as possible like you will live in the future.

So now *apply that to the Christian care for God's good creation.* One day God will renew the whole created order, and according to Romans 8, he will do this by setting over it, as he always intended, his image-bearing creatures. They will reflect God's glory into his world and bring God's saving justice to bear, putting the world to rights and making the desert blossom like the rose. And if we are already in Christ, already indwelt by the Spirit, we cannot say we will wait until God does it in the end. We must be God's agents in bringing, at the very least, signs of that renewal in the present. And that must mean we are called in the present to search out every way in which the present, groaning creation can be set free from at least part of its bondage and experience some of the freedom that comes when the children of God are glorified, because, in Christ and by the Spirit, we already are. To deny a Christian passion for ecological work, for putting the world to rights insofar as we can right now, is to deny either the goodness of creation or the power of God in the resurrection and the Spirit, and quite possibly both.

The "Second Coming"

"Jesus is coming," says the title, so "plant a tree!" The reason that slogan is counterintuitive to the point of being funny is that, for many devout believers, the Second Coming is the point when the world as we know it is done away with and Jesus snatches his own people up to heaven, to live there with him forever. And if that's what's going to happen, why would you plant a tree? Why oil the wheels of a car that's about to drive over a cliff?

But is that what the Second Coming is all about? We will look in a moment at a key passage where many Christians have discovered that sort of doctrine, and I will suggest that it has been sorely misunderstood. But first, let me sketch one or two points of central importance in the New Testament that strongly hint that if we read the Second Coming like that we are missing the point.

A Few Misunderstandings About the Second Coming

In Ephesians 1:10, Paul declares that God's purpose for all eternity was to sum up all things in Christ, in heaven, and on earth. Colossians 1:15–20, one of the most astonishing Christian poems ever written, declares roundly that all things were created in, through, and for Christ, and that all things are reconciled in, through, and for Christ. Neither Ephesians nor Colossians gives any support to the idea that a major part of the created order is destined to be thrown in the trash while redeemed humans, whether in or out of the body, live forever in some other place. Indeed, it's in Colossians 3 that we get a hint of a different way of looking at what is commonly called the Second Coming. Paul says, "You have died, and your life is hidden with Christ in God; when Christ who is our life appears, then you also will appear with him in glory."

Think about the meaning of *appear* for a moment. When we talk of Jesus coming, we make it sound as though he is presently far away; as though, to come, he would have to make a lengthy journey. But *appear* is different. As we find in many passages of the New Testament, Jesus is not far away; he is in heaven, *and heaven is not a place in the sky, but rather God's dimension of what we think of as ordinary reality.* This is an essential feature of biblical cosmology, and the failure to grasp it leaves many Christians puzzled about how to put together the biblical picture of eschatology. The point is that Jesus is presently in God's dimension, that is, heaven; however, heaven is not a place in our space-time continuum, but a different sphere of reality that overlaps and interlocks with our sphere in numerous though mysterious ways. It is as though there were a great invisible curtain hanging across a room, disguising another space that can be integrated with our space; one day the curtain will be pulled back, the two spaces or spheres will be joined forever, and Jesus himself will be the central figure. Now, at that point you could say, if you wanted, that he had come, in the sense that we hadn't been aware of him and now we are. But Colossians speaks of him appearing, and so does 1 John 3:2: "We know that when he appears, we shall be like him, because we shall see him as he is."

I have written at length elsewhere about the "son of man" sayings in Mark 13 and 14 and their parallels, and I don't want to go into them here, except to say this. The picture of the son of man "coming on clouds" in these passages demands to be read in the light of the biblical passage they are quoting, namely Daniel 7; and in Daniel 7 the "one like a son of man" is ascending, not descending. He is coming up to God, not coming back to earth from heaven. I have therefore argued, in line with several other scholars, that these passages are not at all about the Second Coming of Jesus, but rather about his vindication after suffering. I therefore leave them aside not because I don't believe in the Second Coming but because I don't believe these passages teach it.

But the Second Coming itself is presupposed or asserted all over the place in the New Testament. However, it is infrequently integrated with the picture in Romans 8 of the whole creation being renewed, with the new creation born from the womb of the old. There are various passages, though, where something like this is attempted. In 1 Corinthians 15, Paul sketches an extraordinary picture of the coming eschatological scenario, not this time as a mother giving birth but as a great battle between good and evil, a battle that God wins through the victory of Jesus Christ. The whole point of the chapter is that God is going to remake the whole creation; once again Genesis 1, 2, and 3 are in Paul's mind all through, and his focus on the bodily resurrection takes place within the larger scenario where God will defeat the forces of corruption and decay, with the final enemy being death itself. This victory will be the moment of Jesus's "royal appearing," which is the meaning of the Greek word *parousia,* often used these days as if it simply meant "Second Coming" in whatever senses the church understood the phrase. So here in 1 Corinthians we have together the final victory and renewal, the resurrection of the dead, and the royal appearing of Jesus.

Something similar could be said of Revelation 21, though here it is, remarkably, the church, not Jesus, that comes down from heaven. Indeed, Revelation 21 has rightly been hailed as the ultimate rejection of all gnosticism, all attempts to rethink salvation in terms of escape to heaven from earth. On the contrary, the bride of Christ has been waiting in heaven all along and now appears on earth to be married to the bridegroom. Like Romans 8 and 1 Corinthians 15, this is a remarkable picture of continuity and discontinuity. On the one hand, the new creation is genuinely new, to the point that the first heaven and the first earth have passed away. On the other hand, there is clearly strong continuity between the two worlds, since part of God's task in the new world is to wipe away the tears from all eyes. This is a mystery to which we will return.

Before I come to the most contested passage, let me say a word here about those passages, such as 1 Peter, which speak of a salvation being "kept safe for you in the heavens." The point is that heaven is God's space, where God keeps his purposes ready and waiting until it's the right time to bring them out onto the stage of the earth. If I get home late and my wife leaves me a message saying, "Your dinner is in the oven," that doesn't mean she expects me to get into the oven to eat my dinner. It means it's ready and waiting, and when the time comes I can get it out and eat it at the table in the ordinary way.

Our problem here, compounded by the radical misreading of the phrase *kingdom of heaven* to mean a place where God's people go after their death, rather than the reign of heaven, meaning that of God over this earth, is that centuries of supposing that the name of the Christian game was simply to "go to heaven when you die" have left the church, often the very devout church, lingering in a kind of semi-gnosticism without even realizing it, regarding the present world as essentially evil and the purpose of salvation as escape from the world. Not so: the Jesus who taught us to pray, "Your kingdom come . . . on earth as in heaven" also declared that all authority in heaven *and on earth* had been given to him, and Luke's account of the Ascension and what follows indicates well enough that the point of Jesus being in heaven is that he is now enthroned as lord of the world, not that he has gone far away and is merely waiting for us to follow him.

That brings me, again before arriving at the key passage, to two lines that are always quoted in this connection. In John 14, Jesus says that there are many dwelling places in his father's house, and that he's going to prepare a place for his followers. The Greek word for "dwelling places" here is *monai,* the plural of *monē;* and if you look up *monē* in a Greek lexicon you'll find that it doesn't mean "dwelling places" as in "a home where you go and live forever," but rather a lodging house, a place to stay awhile and rest and be refreshed until it's time to continue on your journey. John's Gospel is emphatic about resurrection being the final destiny of God's people; what Jesus is apparently saying is that between death and resurrection there is a place prepared, a place of light and peace and rest, where we can wait in the presence of Jesus until the final day.

That, too, is what Paul means in Philippians 1 when he talks about being "with the king, because that would be far better." So, again, when Jesus says to the brigand in Luke 23 that "you'll be with me in paradise this very day," he isn't talking about the ultimate resurrection life. After all, even Jesus didn't go straight to that, but was well and truly dead and buried and then raised bodily three days later. Where was he in between? *Paradise* is a

Persian word for a beautiful garden that was already used to denote the place where the dead go to be looked after and refreshed until the time of resurrection.

Wonderful Mixed Metaphors in 1 Thessalonians 4

So to 1 Thessalonians 4, where generations have taken Paul's wonderful mixed metaphors as though they were meant literally. After all, in the next chapter Paul declares that the thief is going to come in the night, so the woman will go into labor, so you mustn't get drunk but must stay awake and put on your armor. In chapter 4, writing about a situation where some in the church had died, he has five things to say about what will happen at the end. First, when the last day comes, we will be with those who have died in the Lord. Second, the Christian dead will be raised to new bodily life, and the Christian living will be transformed—essentially the same point he makes at the end of 1 Corinthians 15. Third, this will be a great act of vindication, especially for those Christians who have suffered for the faith. Fourth, this will all happen when Jesus is revealed as king and lord of the whole world, like Caesar but much more so. Fifth, when this happens, the reappearance of Jesus will be like Moses coming down from Mount Sinai with the Law to judge the people.

Now suppose you're Paul, good at mixing your metaphors, and you try to say all those things at once within the biblical cosmology, which uses upstairs/downstairs language for heaven and earth, even though the writers know perfectly well that heaven is not a location in our space-time universe but rather a different kind of space that intersects with ours in complex and interesting ways. How are you going to create this effect? Well, you might try to say that Jesus appearing from heaven will be like Moses reappearing from the cloudy mountaintop after the divine trumpet blast; you might try to put that together with the image from

Daniel about the son of man "coming on clouds," that is, coming upward in vindication; and you might try at the same time to indicate that Jesus's reappearance was like Caesar's "coming" to a town or city, or perhaps returning to Rome after some great victory, with citizens going out to escort him into the city.

And if you are Paul trying to say all those things at once, you might well write something like this: "The Lord himself will descend from heaven with a cry of command, with the call of the archangel and the trump of God; and the dead in Christ will rise first; then we who are still alive will be snatched up on the clouds with them to meet the Lord in the air; and so we shall always be with the Lord. Therefore comfort one another with these words." And Paul would no more think that a whole school of theology might try to combine those into a single literal picture than he would with the bit about the thief and the woman and the sober soldier in the next passage.

I hope you see the point I'm making. It isn't a matter of simply deconstructing the massive "left behind" theology that has been so powerful in North America in particular, though we must do that if we are to think biblically. We must focus on one element in particular. The word *parousia,* "royal appearing," was regularly used to describe Caesar's "coming" or "royal appearing" when visiting a city, or when returning home to Rome. And what happened at such a *parousia* was that the leading citizens would go out to meet him, the technical term for such a meeting being *apantçsis,* the word Paul uses here for "meeting," as in "meeting the Lord in the air." But when the citizens went out to meet Caesar, they didn't stay there in the countryside. They didn't have a picnic in the fields and then bid him farewell; they went out *to escort their Lord royally into their city.* In other words, Paul's picture must not be pressed into the nonbiblical image of the Second Coming according to which Jesus is "coming back to take us home"—swooping down, scooping up his people, and zooming back to heaven with them, away from the wicked earth forever. As Revelation makes clear in several passages, with echoes in other New Testament books, the point is that Jesus will reign on the earth, and at his royal appearing the faithful will go to meet him, like the disciples on the road to Jerusalem only now in full-blooded triumph, and escort him back into the world that is rightfully his and that he comes to claim, to judge, to rule with healing and wise sovereignty.

"When Christ shall come," we sing in a favorite hymn, "with shout of acclamation, and take me home, what joy shall fill my heart." What we ought to sing is, "When Christ shall come, with shout of acclamation, *and heal his world,* what joy shall fill my heart." In the New Testament the Second Coming is not the point at which Jesus snatches people up, away from the earth, to live forever with him somewhere else, but the point at which he returns to reign not only in heaven but upon the earth. After all, the risen Jesus in Matthew 28 declares that "all authority in heaven and on earth has been given to me," and makes that the basis for his commission to his disciples.

What, then, of the frequently quoted line, "My kingdom is not of this world" (John 18:36)? Well, the main thing is that the translation is misleading. The Greek is *ek tou kosmou toutou,* which means "from this world," not "of." Jesus's kingdom is not *from* this world, that is, it doesn't originate in the present world order; if it did, as he goes on to say, his followers would use violence. But there is no question, in John's Gospel, that Jesus's kingdom is, so to speak, *for* this world. It comes from heaven but is destined for earth, or rather for the new heavens and new earth spoken of elsewhere in the New Testament. That is why, as the Judaean leaders declare, if Pilate lets him go, he is no friend of Caesar. There is, I have to say, a terrible irony about people in my country a century or so ago, and others in more recent times, loudly insisting that Jesus's kingdom is purely spiritual and then claiming the freedom to use violence to advance their own kingdoms. But that is a topic for another occasion.

There is much more I could say about the Second Coming, but you will have to look it up in my various books, particularly *The Resurrection of the Son of God, Surprised by Hope,* and chapter 11 of *Paul and the Faithfulness of God.*

2 Peter 3

My final subject is a tricky little passage in 2 Peter 3. I do not think we can be absolutely sure what this means, though we can have a good try. I am going to quote from the relevant passage in my large book on the resurrection (*The Resurrection of the Son of God,* pp. 462–63).

The critical moment here, upon which seems to hinge the worldview of the whole, is verse 10: "But the Lord's day will come like a thief. On that day the heavens will pass away with a great rushing sound, the elements will be dissolved in fire, and the earth and all the works on it will be disclosed." Is the writer saying that creation as a whole is to be thrown away and a new one, freshly made, to take its place? So it would seem if the verse were to end "will be burned up," as in the Authorized Version and Revised Standard Version. That could imply a dualistic worldview in which creation itself was irremediably evil, which seems ruled out by the insistence on its being divinely made, or a Stoic worldview in which the present world would dissolve into fire and be reborn, phoenix-like, from the ashes, which seems ruled out by the fact that the underlying story is not one of an endless cycle, as in Stoicism, but of a linear movement of history, as in Judaism, moving forward toward judgment and new creation. What is going on in this text, and what view does it offer of the future world, and of humans within it?

The translation "will be burned up" depends on the variant readings of a few manuscripts. Most of the best witnesses have *heurethesetai,* "will be found." Until recently it was thought that this was quite unintelligible, but more recently commentators have pointed out the use of "find" in the sense of "being found out," in a setting of eschatological judgment, in Jewish texts and elsewhere in the New Testament, including Paul and the Gospels. Various possible nuances of meaning emerge from this, of which one stands out: that the writer wishes to stress continuity within discontinuity, a continuity in which the new world, and the new people who are to inhabit it, emerge tested, tried, and purified from the crucible of suffering. If something like this is plausible, then the worldview we glimpse here is not that of the dualist who hopes for creation to be abolished, but of one who, while continuing to believe in the goodness of creation, sees that the only way to the fulfillment of the creator's longing for a justice and goodness to replace the present evil is for a process of fire not simply to consume but also to purge, to "discover" the deepest truth of the good creation beneath the overlayering of corruption and wickedness. The second letter of Peter may thus offer further witness to an early Christian eschatology not far removed from that of Paul in 1 Corinthians.

I WISH I HAD time to explore the theme of creation and new creation through both the New Testament and the second-century fathers. It is striking how the earliest Christians, like mainstream rabbis of the period, clung to the twin doctrines of creation and judgment: God made the world and made it good, and one day he will come and sort it all out. Take away the goodness of creation, and you have a judgment where the world is thrown away as so much garbage, leaving us sitting on a disembodied cloud playing disembodied harps. Take away judgment, and you have this world rumbling on with no hope except the pantheist one of endless cycles of being and history. Put creation and judgment together, and you get new heavens and new earth, created not *ex nihilo* but *ex vetere,* not out of nothing but out of the old one, the existing one.

And the model for that is of course the resurrection of Jesus, who didn't leave his body behind in the tomb and grow a new one but whose body, dead and buried, was raised to life three days later and recognized by the marks left by the nails and the spear. There is a whole world of eschatological understanding in the resurrection narratives, not least John's insistence that Easter day is the first day of the week; John, so rooted in creational theology, knows exactly what he is doing. Easter is the beginning of God's new creation. We don't have to wait. It has already burst in. And the whole point of John 20 and 21 is that we who believe in Jesus are to become, in the power of his spirit, not only beneficiaries of that new creation but also agents.

I end with an extraordinary verse, 1 Corinthians 15:58: "So, my dear family, be firmly unshakable, always full to overflowing with the Lord's work. In the Lord, as you know, the work you're doing will not be worthless." Now what is that exhortation doing at the end of a chapter on resurrection? If we were to take the normal Western view of life after death, a long chapter on resurrection might end up with something like this: "Therefore, my beloved, lift up your head and wait for the wonderful hope that is coming to you eventually." But for Paul, as is clear throughout 1 Corinthians, *the resurrection means that what you do in the present matters into God's future.* That is so for ethics, not least sexual ethics, as in 1 Corinthians 6. But it is also so for everything else. The resurrection, God's recreation of his wonderful world, which began with the resurrection of Jesus and continues mysteriously as God's people live in the risen Christ and in the power of his spirit, means that what we do in Christ and by the Spirit in the present is not wasted. It will last and be enhanced in God's new world.

I have no idea precisely what this means. I do not know how the painting an artist paints today in prayer and wisdom will find a place in God's new world. I don't know what musical instruments we will have to play Bach, though I'm sure Bach's music

will be there. I don't know how my planting a tree today will relate to the wonderful trees that will be in God's recreated world.
I don't know how my work for justice for the poor, for remission
of global debts, will reappear in that new world. But I know
that God's new world of justice and joy, of hope for the whole
earth, was launched when Jesus came out of the tomb on Easter
morning; I know he calls me and you to live in him and by the
power of his spirit, and so to be new-creation people here and
now, giving birth to signs and symbols of the kingdom on earth
as in heaven. The resurrection of Jesus and the gift of the Spirit
mean that we are called to bring forth real and effective signs of
God's renewed creation even in the midst of the present age. Not
to do so is at best to put ourselves in the position of those Second
Temple Jews who believed they had to wait passively for God to
act—when God *has* acted in Jesus to inaugurate his kingdom on
earth as in heaven. At worst, not to bring forth works and signs of
renewal in God's creation is to collude, as gnosticism always does,
with the forces of sin and death.

This doesn't mean that we are called to build the kingdom by
our own efforts, or even with the help of the Spirit. The final
kingdom, when it comes, will be the free gift of God, a massive act of grace and new creation. But we are called to build *for*
the kingdom. Like craftsmen working on a great cathedral, we
have each been given instructions about the particular stone we
are to spend our lives carving, without knowing or being able
to guess where it will take its place within the grand design. We
are assured, by the words of Paul and by Jesus's resurrection as
the launch of that new creation, that the work we do *is not in
vain*. That says it all. That is the mandate we need for every act
of justice and mercy, every program of ecology, every effort to
reflect God's wise stewardly image into his creation. In the new
creation, the ancient human mandate to look after the garden
is dramatically reaffirmed—another point we could draw out of

John 20 were there time. The resurrection of Jesus is the reaffirmation of the goodness of creation, and the gift of the Spirit is
there to make us the fully human beings we were supposed to be,
precisely so that we can fulfill that mandate at last. What are we
waiting for? Jesus is coming. Let's go and plant those trees.

6

9/11, Tsunamis, and the
New Problem of Evil

IN THE NEW heaven and new earth, according to Revelation 21, there will be no more sea. Many people feel disappointed by this. The sea is a perennial delight, at least for those who don't have to make a living on it. What is going on? The sea is part of the original creation, of which God says that it is "very good." But already by the story of Noah the flood poses a threat to the creation, with Noah and his floating zoo rescued by God's grace. From within the good creation come forces of chaos, harnessed to enact God's judgment. We then find Moses and the Israelites standing at the edge of the sea, chased by the Egyptians and at their wits' end. God makes a way through the sea to rescue his people and judge the pagan world, like the Noah story but in a new mode. As later poets look back on this decisive moment in the story of God's people, they celebrate it in terms of the old creation myths: the waters saw YHWH and were afraid, and they went backward.

But then, in a passage that greatly influenced early Christianity, we find the vision of Daniel 7, where the monsters who make war upon the people of the saints of the most high come up out of the sea. The sea has become a dark, fearsome, threatening place from which evil emerges, threatening God's people like a giant tidal wave threatening those who live near the coast. For the people of ancient Israel, who were not for the most part seafarers, the sea came to represent evil and chaos, the dark powers that might do to God's people what the flood had done to the whole world, unless God rescues them as he rescued Noah. This sets the biblical context for reflection on the Indian Ocean earthquake and tsunami of December 26, 2004.

It may indeed be that one of the reasons we love the sea is because, like watching a horror movie, we can observe its power and relentless energy from a safe distance, or if we go sailing on or swimming in it, we can use its energy without being engulfed by it. I suspect there are plenty of Ph.D. theses already written on what goes on psychologically when we do this, but I haven't read them. We would, however, find our delight turning quickly to horror if, as we stood watching the waves roll in, a tsunami were suddenly to come crashing down on us, just as our thrill at watching a gangster movie would turn to screaming panic if a couple of thugs, armed to the teeth, came out of the screen and threatened us personally as we sat innocently in the cinema. The sea and the movie, seen from a safe distance, can be a way of saying to ourselves that, yes, evil may well exist; there may be chaos out there, but thank goodness, we are not immediately threatened by it. And perhaps this is also saying that, yes, evil may also exist inside us: there may be forces of evil and chaos deep inside us of which we are at best only subliminally aware, but we are in control—the sea wall will hold, and the cops will get the gangsters in the end.

In movies of the last decade or two, things may not work out so well, which may tell us something about how we now perceive evil both in the world and in ourselves. And when a real tsunami wipes out a quarter of a million people in a day, we are forced to take stock of everything in a new way. We are forced to think in fresh ways about the problem of evil, maybe even to see that there is what I am calling here "a new problem of evil."

What do I mean by this? A number of years ago, President George W. Bush declared that there is an "axis of evil" out there and that we have to find the evil people and stop them from doing more evil. Tony Blair said in 2002, ambitiously, that the aim of his government must be nothing short of ridding the world of evil. The press shout "evil" at the faces of terrorists and child-murderers. And the odd thing about this new concentration on evil is that it seems to have taken many people, not least politicians and the media, by surprise. Of course they would say that there has always been evil, but it seems to have come home to the Western world in a new way. Older discussions of evil tended to be more abstract, with so-called natural evil (represented by the tidal wave) and so-called moral evil (represented by the gangsters). But the tsunami and the gangsters came out of the theory books and attacked. Just as in the previous generation Auschwitz posed the problem in a new way, September 11, 2001, kick-started a fresh wave of discussion about what evil is, where it comes from, how to understand it, what it does to your worldview whether you're a Christian or an atheist or anything else, and not least, what if anything can be done about it. And, though terrorism and tsunamis are not the same kind of things, the events of December 26, 2004, have added their own dimension to the problem.

From the Christian point of view, there will be no more sea in that sense in the new heavens and new earth. We are committed, within the worldview generated by the gospel of Jesus, to affirming that evil will finally be conquered, will be done away with. But understanding why it's still there as it is, and how God has dealt with it and will deal with it, how the cross of Jesus has anything to do with that, how it affects us here and now, and what we can do here and now to be part of God's victory over evil—all these are deep and dark mysteries that the sudden flurry of new interest in evil opens up as questions. What I want to do here is to sketch out the areas where we should be urgently thinking and praying.

An Inadequate View of the Problem of Evil

The older ways of talking about evil tended to pose the puzzle as a metaphysical or theological conundrum. If there is a god, and if he is a good, wise, and supremely powerful god, why is there such a thing as evil? Even if you're an atheist, you face the problem: Is this world a sick joke that contains some things to make us think it's a wonderful place and other things to make us think it's an awful place, or what? You could, of course, call this the problem of good rather than the problem of evil: If the world is the chance assembly of accidental phenomena, why is there so much that we want to praise and celebrate? Why is there beauty, love, and laughter?

The problem of evil in its present metaphysical form has been around for at least two and a half centuries. The Lisbon earthquake on All Saints' Day 1755 shattered the easy optimism, the straightforward natural theology, of the previous generation. The wrestling of the great Enlightenment thinkers—Voltaire, Rousseau, Kant, and Hegel—can be understood as ways of coping with evil. And when we come to Marx and Nietzsche, and to the twentieth-century thinkers, not least Jewish thinkers who wrestled with the question of meaning following the Holocaust, we find a continuous thread of philosophical attempts to talk about the world as a whole and about evil within it.

Unfortunately, the line of thought that has emerged at a popular level is less satisfactory. I refer to the doctrine of progress. Everything is moving toward a better, fuller, more perfect end. If there has to be suffering on the way, so be it; omelets are made from broken eggs.

This belief in progress, which you find in poets such as Keats, was in the air in the pantheism of the Romantic movement and was given an enormous boost by the popularization of Charles Darwin's ideas and their application to fields considerably more diverse than the study of birds and mammals. The heady combination of technological achievement, medical advances, Romantic pantheism, Hegelian progressive idealism, and social Darwinism created a climate of thought in which, to this day, a great many people, not least in public life, have lived and moved. The world is improving! It's our article of faith. Things are getting better. Evil is on the decline. Progress is winning. This belief in progress has overcome many challenges, and it has remarkably survived and flourishes. The First World War shook the old liberal idealism; Auschwitz shattered the idea of Western civilization as a place of nobility, virtue, and humanizing reason. Yet, despite everything, people still suppose that the world is basically a good place and that its problems are more or less soluble by technology, education, "development" in the sense of "Westernization," and the application or imposition of Western democracy and capitalism.

This state of affairs has led to three things that I see as characterizing the new problem of evil. First, we ignore evil when it doesn't hit us in the face. You see this in the philosophy, psychology, and politics of the last century, but you don't need to look that far back. Western politicians knew perfectly well that al-Qaeda was a danger, but nobody took it seriously until it was too late. Countries bordering the Indian Ocean knew about tsunamis but hadn't bothered to install early warning systems. We all know that Third World debt is a massive sore on the conscience of the world, but our politicians don't want to take it too seriously, because from our point of view the world is progressing reasonably well and we don't want to rock the economic boat—or upset powerful interests. We all know that sexual licentiousness creates terrible unhappiness in families and individual lives, but we live in the twenty-first century, and we don't want to say that adultery is wrong. We live in a world where politicians, media pundits, economists, and even, alas, some late-blooming liberal theologians, speak as if humankind is basically all right, the world is basically all right, and there's nothing we should make a fuss about.

So, first, we ignore evil except when it hits us in the face. Second, we are surprised by evil when it does. We like to think of small English towns as pleasant, safe places and were shocked to the core a few years ago when two little girls were murdered by someone they obviously knew and trusted. We have no categories to cope with that, nor do we have categories to cope with evils such as renewed tribalism and genocide in Africa. We like to fool ourselves that the world is basically all right, now that so many countries are either democratic or moving that way and globalization has in theory enabled us to do so much, to profit so much, to know so much. Then we are puzzled and shocked by the human tsunami, the great wave of refugees and those seeking asylum. Terrorism takes us by surprise, since we are used to imagining that all serious questions should be settled around a discussion table and are puzzled that some people still think they need to use more drastic methods of getting their point across. And ultimately, we are shocked again and again by death. We ignore evil when it doesn't hit us in the face, and so we are shocked and puzzled when it does.

As a result, we react in immature and dangerous ways. The reaction in America and Britain to the events of September 11, 2001, was a knee-jerk, unthinking, immature lashing out. Don't misunderstand me. The terrorist actions of al-Qaeda were and are unmitigatedly evil. But the astonishing naïveté which decreed that America as a whole was a pure, innocent victim, so that the world could be neatly divided up into evil people (particularly Arabs) and good people (particularly Americans and Israelis), with the latter having a responsibility to punish the former, thus justifying the wars in Afghanistan and Iraq, is a large-scale example of what I'm talking about—just as it is immature and naïve to suggest the mirror image of this view, namely that the Western world is guilty in all respects and that all protestors and terrorists are therefore completely justified in what they do.

The Western world, then, has not been able to cope with evil from within its modernistic beliefs. Postmodernity doesn't help here, because it remorselessly highlights the problem of evil, while avoiding any return to a classic doctrine of original sin by denying that there is really anybody there in the first place. Humans themselves deconstruct; you can't escape evil in postmodernity, but there is nobody to take the blame. There is no moral dignity left. To shoulder responsibility is the last virtue left open to those who have forsworn all other kinds; to have even that disallowed is to reduce human beings to mere ciphers. Most of us, not least genuine victims of crime and abuse, find that both counterintuitive and disgusting. Furthermore, postmodernity's analysis of evil allows for no redemption. There is no way out, no chance of repentance and restoration, no way back to the solid ground of truth from the quicksands of deconstruction. Postmodernity may be correct in saying that evil is real, powerful, and important, but it gives us no real clue as to what we should do about it. It is therefore vital that we look elsewhere and broaden the categories of the problem from the shallow modernist puzzles on the one hand and the nihilistic deconstructive analyses on the other. This sends us back to the Bible itself. What has it got to say about all this?

The Biblical Response to the Problem of Evil

There are three things to say by way of introduction to this sketch of a biblical response to the new problem of evil. First, there are no easy answers, in scripture or elsewhere. If we think we've solved the problem of evil, that just shows we haven't understood it. We cannot say, with that dreadful hymn, that "all our pain is good." That induces a moral chaos worse than the one created by Job's comforters. Nor can we say that evil is good after all because it provides a context for moral effort and even heroism, as though we could get God off the hook by making the world a theater where God sets up little plays to give his characters a chance to show how virtuous they really are. That is trivializing to the point of blasphemy. So, first, no easy answers.

Second, the line between good and evil does not lie between "us" and "them," between the West and the rest, between Left and Right, between rich and poor. That fateful line runs down the middle of each of us, every human society, every individual. This is not to say that all humans, and all societies, are equally good or bad; far from it. Merely that we are all infected and that all easy attempts to see the problem in terms of "us" and "them" are fatally flawed.

Third, though there is indeed a radical difference in kind between the problem posed by 9/11 and the problem posed by the tsunami—no terrorists or wicked governments were responsible for the tsunami; a tectonic plate's just got to do what a tectonic plate's got to do—they pose, together, the great question that the Bible was in fact written to address, even though we cannot fully understand the answers until the end.

What help can we find in the Bible? We have already seen that the sea, both in itself and in its symbolic significance, is simultaneously part of God's good creation and a continuing source of chaos and terror. We should also note, to glance ahead to the very end of the story, that God declares throughout scripture that he is going to put the world to rights at the last, even though this will involve, in Haggai's phrase, giving both heaven and earth one last great shake to sort everything out. Scripture seems to be trying to say that creation is good but incomplete, and human evil has somehow stalled the project of creation in its incomplete mode, so that humans need to be put right and the world needs a good shake. The biblical imagery of judgment insists over and over that God will put the world to rights and wipe every tear from every eye.

But how is this to come about? Here I believe we have gone astray. We are accustomed to seeing what we call "the problem of evil" in terms of the philosophical puzzle of how to justify a good, wise, and powerful God in the face of continuing evil. We have therefore gone to the book of Job as the one part of scripture that addresses the problem in broadly those terms. But we have failed to see that the rest of scripture addresses the problem of evil in a different way. After Noah comes the debacle of the Tower of Babel, and immediately after that, God calls Abraham and declares that in him all the world will be blessed. The curse of evil at every level is to be undone. The Israelite prophets and poets who saw this most clearly spoke of creation healed, of the wolf and the lamb lying down together, of swords beaten into plowshares, of shrubs and fruit trees replacing the thorns, thistles, and briars that had defaced the garden.

And they spoke of evil itself in a variety of ways. They described evil in terms of wicked nations oppressing God's people, wicked rulers oppressing the poor, wicked people within society using money or power to oppress one another. They spoke of evil in terms of the failure of human beings in general, and of Israel in particular, to live as image-bearing human beings should, to live as the redeemed people of God ought to do. They thus spoke of evil at the level both of individuals and of society; of evil as a problem infecting all, including God's people; of evil in terms of both human wickedness and demonic activity, of structural injustice and systemic oppression and violence, particularly that of the great empires of the world.

The opening chapters of Genesis make all this extremely clear. But then, in Abraham, God declares, as an act of sovereign grace following the word and act of judgment, that a new way has opened up, a way by which the original purpose of blessing for humankind and creation can be taken forward. From within the story we already know that this is going to be enormously costly for God. The loneliness of God looking for his partners, Adam and

Eve, in the garden; the grief of God before the flood; the head-shaking exasperation of God at Babel; the frustration of God at Abraham's vacillating half-obedience—all these God will have to continue to put up with. There will be numerous further acts of judgment as well as mercy as the story unfolds. But unfold it will. The overarching picture is of the sovereign creator God who continues to work within his world—that's the key—until blessing replaces curse, homecoming replaces exile, olive branches appear after the flood, and a new family is created in which the scattered languages can be reunited. That narrative forms the frame for the canonical Old Testament. It's not a matter of God as a puppet master, pulling the strings from afar. Rather God loves his world so much that, faced with evil within it, he works within the world, despite the horrible ambiguities that result.

The body of the Old Testament, from this point on, carries—and the writers know it carries—the deeply ambiguous story of Abraham's family, the people through whom God's solution was taken forward, which included people who were themselves part of the problem. The Exodus gives the major paradigmatic Jewish answer to the question "What is God doing about evil?" Evil comes here in the shape of the wicked powerful empire oppressing the enslaved Israelites. But when the people are freed, they behave in a thoroughly pagan manner, as they continue to do in the deeply ambiguous entry into the land, in the period of the Judges, and then under the monarchy. David, a man after God's own heart, is himself deeply flawed. Exile threatens, and then, like a great tsunami, Babylon comes and sweeps away the temple where YHWH had lived in the midst of his people. And though the people return two generations later, all is not well. Pagan empires still lord it over them. The sea monsters still attack. God's people still cry for YHWH to come and redeem them, to put the world to rights once and for all.

At the heart of the greatest prophetic book of all, Isaiah, we find the servant of YHWH, who does two things at once: he represents Israel within the purposes of God, and he embodies God's rescue operation for Israel and the world. Indeed, it is immediately after his suffering and death in chapter 53 that we have the word of a new covenant in chapter 54 and a new creation in chapter 55. Somehow, the prophet is saying, the people of Israel, the bearers of the solution, have become part of the problem, but as God had determined to work from within his world to rescue his world by calling Israel in the first place, so he has determined to work from within Israel to rescue Israel by calling this royal yet suffering figure, by equipping him with his own spirit and allowing the worst the world can do to fall upon him. If you want to understand God's justice in an unjust world, says Isaiah, this is where you must look. God's justice is saving, healing, restorative, because the God to whom justice belongs has yet to complete his original plan for creation, and his justice not only restores balance to a world out of kilter but brings to glorious fruition the creation teeming with life and possibility that he made in the first place.

Somehow Isaiah has so redefined the broader problem of evil, of the injustice of the world and the justice of the one creator God, that we now see it not as a philosopher's puzzle requiring explanation but as the tragedy of creation requiring a fresh intervention from the sovereign creator God, as well as the tragedy of Israel requiring renewal from the sovereign covenant God. And to our amazement and horror, we see this renewal come into focus in the suffering and death of the servant. Sharing the fate of Israel in exile, the exile that we know from Genesis 3 onward is closely aligned with death itself, as the servant bears the sin of the many. He embodies the covenant faithfulness and restorative justice of the sovereign God, and by his stripes we are healed.

Central to the Old Testament picture of God's justice in an unjust world, then, is the picture of God's faithfulness to unfaithful Israel; and central to that picture is YHWH's servant, an individual who stands over against Israel, taking its fate upon himself so that Israel may be rescued from exile and the human race proceed at last toward the new creation, in which thorns and thistles will be replaced by cypress and myrtle, dust and death by new life, the chaos of the sea by the victory of the creator. Within the larger canonical context, we might see the entire book of Job alongside Isaiah as an anticipation of the harrowing scene in Gethsemane, where the comforters again fail and even creation goes dark as the monsters close in around the innocent figure who asks what it's all about.

That leads us to the heart of the matter. The moment when the sinfulness of humankind grieved God to his heart, the moment when the servant was despised and rejected, the moment when Job asked God why it had to be that way, all come together when the Son of Man kneels, lonely and afraid, before going to face the might of the beasts that had come up out of the sea. The story of Gethsemane and the cross present themselves in the New Testament as the strange, dark conclusion to the story of what God does about evil, of what happens to God's justice when it takes human flesh, when it gets its feet muddy in the garden and its hands bloody on the cross. The multiple ambiguities of God's actions in the world come together in the story of Jesus.

The Cross and Evil

Theologies of the cross, of atonement, have not in my view grappled sufficiently with the larger problem of evil as normally conceived. Conversely, those who have written about the problem of evil in philosophical theology have not normally grappled sufficiently with the cross as part of both the analysis and the solution. The two have been held apart, with the problem of evil conceived simply in terms of how a good and powerful God could allow evil into the world in the first place and the atonement seen in terms of personal forgiveness. Much modern Christian thought has accepted the framework offered by the Enlightenment, in

which the Christian faith *rescues* people *from* the evil world, en-
suring them forgiveness in the present and heaven hereafter. The
Enlightenment-based wider world then accepted that evaluation
of the Christian faith—not surprisingly, since it was driving it
in the first place—and so has not thought it necessary to factor Christian theology into its own discussions of the problem of
evil. How, after all, does a hymn like "There Is a Green Hill Far
Away" have anything at all to say to a world dumbstruck in horror at the First World War, Auschwitz, Hiroshima, 9/11?

With this in mind, we need to reread the Gospels as what they
are. People often observe that there is not that much atonement
theology in the Gospels. Mark's theology of the cross is often
reduced to one key verse, 10:45, which speaks of the Son of Man
coming "to give his life a ransom for many." The Lord's Supper
gives hints of an atonement theology, and the crucifixion narratives, especially in their evocation of biblical allusions, provide
further elements. But for the most part the Gospels, as read within
the mainstream tradition of scholarship and church life, have little to contribute, except as backdrop to an atonement theology
grounded elsewhere, in Paul, Hebrews, and 1 Peter.

But when we read the Gospels in a more holistic fashion, we
find that they tell a double story, drawing together the themes I
have described. They tell the story of how the evil in the world—
political, social, personal, moral, and emotional—reached its
height; they tell how God's long-term plan for Israel—and for
himself!—finally came to its climax. And they tell both of these
stories in the story of how Jesus of Nazareth announced God's
kingdom and went to his violent death. The Gospels, read in this
way, offer us a richer theology of atonement than we are used to
and also a deeper understanding of the problem of evil and what
must be done about it in our own day. The Gospels have more to
say about terrorism and tsunamis than we might imagine.

Watch as they tell how all the varied forces of evil are involved
in putting Jesus on the cross. They tell how the political powers

of the world reached their full, arrogant height: Rome and Herod stand in the near distance, as does Caiaphas and his corrupt Jerusalem regime. All three come into focus as the cross comes closer. Thus the Gospels tell a story of corruption within Israel itself. The Pharisees offer a hard-edged interpretation of Torah that excludes the eruption of God's kingdom through Jesus. The revolutionaries try to get in on the inauguration of God's kingdom, but they, the terrorists of their day, try to fight violence with violence and so merely collude with evil. The death of Jesus, when it comes, is the work not only of the pagan nations but of Israel, which has reduced itself to saying that it has no king but Caesar.

The Gospels also tell the story of deeper, darker demonic forces that operate at a suprapersonal level. These forces operate through all the human elements I've mentioned but cannot be reduced to them. The shrieking demons that yell at Jesus, that rush at him out of the tombs, are signs that a battle has been joined at more than a personal level. The stormy sea, the miniature but deadly tsunamis on Galilee, evoke ancient Israelite imagery of an evil that is more than the sum total of present wrongdoing and woe. "The power of darkness" to which Jesus alludes immediately before his betrayal suggests that on that night evil was given free rein to do its worst, with the soldiers, the betrayer, the muddled disciples, and the corrupt court as mere instruments. The mocking bystanders as Jesus hangs on the cross ("If you are the son of God . . .") echo the taunting, tempting voice that had whispered in the desert. The power of death, the ultimate denial of the goodness of creation, speaks of a force of destruction, of antiworld, anti-God power being allowed to do its worst.

The Gospels tell this story in order to say that the tortured young Jewish prophet hanging on the cross was the point where evil, including the violence of terror and the nonhuman forces that work through creation, had become truly and fully and totally itself. The Gospels tell the story of the *downward spiral* of evil. One thing leads to another; the remedy against evil has itself the germ of evil within it, so that its attempt to put things right merely produces second-order evil. And so on. Judas's betrayal and Peter's denial are simply final twists of this story, with the casual injustice of Caiaphas and Pilate and the mocking of the crowds at the cross tying all the ends together.

Once we learn to read the Gospels in this holistic fashion, we hear them telling us that the death of Jesus is the result both of the major *political* evil of the world, the power games that the world still plays, *and* of the dark, accusing forces that stand behind those human and societal structures, forces that accuse creation itself of being evil, and so try to destroy it while its creator longs to redeem it. The Gospels recount Jesus's death as a story of how the downward spiral of evil finally hit bottom with the violent and bloody execution of *this* man, this prophet who had announced God's kingdom.

In the story of Jesus, and particularly his death, cosmic and global evil, in its suprapersonal as well as personal forms, is met by the sovereign, saving love of Israel's God, YHWH, the creator of the world. The Gospels intentionally draw the Old Testament narrative to its climax, framing that narrative as the story of God's strange and dark solution to the problem of evil from Genesis 3 on. What the Gospels offer is not a philosophical explanation of evil—what it is or why it's there—but the story of an *event* in which the living God *deals with it*. Like the Exodus from Egypt or the return from Babylon, only now with fully cosmic reach, God has rescued his people from the dark powers of chaos. The sea monsters have done their worst, but God has vindicated his people and put creation to rights.

And he has done so *through* the suffering of Israel's representative, the Messiah. This is what it looks like when YHWH says, as in Exodus 3:7-8, "I have observed the misery of my people . . . I have heard their cry . . . and I have come down to deliver them." This is what it looks like when YHWH says "Here is my servant" (Isaiah 42:1). As Isaiah says later (chapter 59), it was no messenger, no angel, but his own presence that saved them; in all their affliction, he was afflicted. God chose the appropriate and necessarily deeply ambiguous route of acting from *within* his creation, from *within* his chosen people, to take the full force of evil upon himself and so exhaust it. And the result is that the covenant is renewed. Sins are forgiven; the long night of sorrow, exile, and death is over and the new day has dawned. New creation has begun, the new world in which violence will be overcome and the sea will be no more.

The Gospels thus tell a story unique in the world's great literature, religious theories, and philosophies: the story of the creator God taking responsibility for what's happened to creation, bearing the weight of its problems on his own shoulders. As Sydney Carter put it in one of his finest songs, "It's God they ought to crucify, instead of you and me." Or as one old evangelistic tract put it, the nations of the world got together to pronounce judgment on God for all the evils in the world, only to realize with a shock that God had already served his sentence. The tidal wave of evil crashed over the head of God himself. The spear went into his side like a plane crashing into a great building. God has been there. He has taken the weight of the world's evil on his own shoulders. This is not an explanation. It is not a philosophical conclusion. It is an event in which, as we gaze on in horror, we may perhaps glimpse God's presence in the deepest darkness of our world, God's strange unlooked-for victory over the evil of our world; then, and only then, we may glimpse God's vocation for us to work with him on a solution to the new problem of evil.

We Are Called to Implement God's Victory in the World

I spoke earlier of the shallow analysis of evil and the immature reactions it produces. It is fascinating that the best known of the

Gospel atonement passages occurs in the context of a sharp saying of Jesus about the nature of political power and its subversion by Gospel events. The request of James and John that they sit on either side of Jesus when he comes into his kingly power is a political question that receives a political answer: earthly rulers lord it over their subjects, but it must not be so among you. Rather those who are great must be the servants, and those who are chief must be slaves of all, because the Son of Man came not to be served but to serve and to give his life as a ransom for many (Mark 10:35–45). This evocation of Isaiah 53, in a manner entirely true to the original context, sits in the middle of the *political analysis of empire* and subverts that empire, based as it is on violence, by showing how the traditions of Israel, the people through whom God would address and solve the problem of the world's evil, converge on a figure who takes all that Babylon, all that Rome, can do to him and exhausts and defeats it.

We find the same point in Luke 9:54, where once more James and John want to do things in the world's way, calling down fire from heaven on their enemies. Jesus's rebuke to them (9:55) is directly cognate with the "Father, forgive them" that he gasps out on the cross.

What, then, is the result? The call of the Gospel is for the church to *implement* the victory of God in the world. The cross is not just an example to be followed; it is an achievement to be worked out, put into practice. But it is an example nonetheless, because it is the exemplar, the template, the model, for what God now wants to do, by his spirit, in the world, through his people. It is the start of the process of redemption, in which suffering and martyrdom are the paradoxical means by which victory is won. The suffering love of God, lived out again by the Spirit in the lives of God's people, is the God-given answer to the evils of the world.

But what if (someone will ask)—what if the people who now bear the solution become themselves part of the problem, as happened before? Yes, that is a danger, and it must be addressed. The church is never more at risk than when it sees itself merely as the solution bearer and forgets that every day it must say, "Lord, have mercy on me, a sinner," and allow that confession to work its way into genuine humility even as it stands boldly before the world and its crazy empires. In particular, it is a problem when a Christian empire seeks to impose its will, dualistically, on the world, by labeling other parts of the world evil while seeing itself as the avenging army of God. That is more or less exactly what Jesus found in the Israel of his day, what he saw in James and John. The cross was and is a call to a different vocation, a new way of dealing with evil, ultimately a new vision of God.

What, after all, would it look like if the true God came to deal with evil? Would he come in a blaze of glory, in a pillar of cloud and fire, surrounded by legions of angels? Jesus of Nazareth took the total risk of speaking and acting as if the answer to the question were this: when the true God comes back to deal with evil, he will look like a young Jewish prophet journeying to Jerusalem at Passover time, celebrating the kingdom, confronting corrupt authorities, feasting with his friends, succumbing in prayer and agony to a cruel and unjust fate, taking upon himself the weight of Israel's sin, the world's sin, Evil with a capital *E*. When we look at Jesus in this way, we discover that the cross has become for us the new temple, the place where we go to meet the true God and know him as savior and redeemer. The cross becomes the place of pilgrimage where we stand and gaze at what was done for each of us. The cross becomes the only sign by which we go to address the wickedness of the world. The cross signifies that the pagan empire, symbolized in the might and power of brute force, has been decisively challenged by the power of love—and that this decisive challenge will win the day. That is the Christian answer to the problems summed up for us in 9/11.

What, then, about the tsunami? There is no straightforward answer, but there are small clues. We are not to suppose that the current world is the way God intends it to be at the last. Some serious thinkers, including some contemporary physicists, link the convulsions that still happen in the world to evil perpetrated by humans, and it is indeed reasonable to probe for deeper connections than modernist science has imagined between human behavior and the total environment of our world, including tectonic plates. But I find it somewhat easier to suppose that the project of creation, the good world that God made at the beginning, was supposed to go forward under the wise stewardship of the human race as God's vicegerents and image bearers, and that when the human race turned to worship creation instead of God, the project could not proceed in the intended manner but instead bore thorns and thistles, volcanoes and tsunamis, the terrifying wrath of the creation that we humans had treated as if it were divine.

When we go to the Gospels for help, we should listen to what they actually say. Matthew tells the story of God-with-us, Emmanuel—with us in the middle of the swirling, raging waters, asleep in the boat on the lake, vulnerable to the screams of the demoniacs and the plots of the Pharisees, undermined by his own associates and finally hunted down by the chief priests and handed over to the imperial authorities. Matthew would forbid us to ask the question about the tsunami in terms of a God who sits upstairs and pulls the puppet strings to make things happen, or not, as the case may be, down here. We must tell the story in terms of the God who was with his people in the midst of the mighty waters, the God who was swept off his feet and out to sea, the God who lost his parents and family, the God who was crushed under falling concrete and buried in mud.

And then we have to learn to also tell the story in terms of the God who rescued others while not saving himself, the God who worked night and day to recover bodies and some still alive, the God who rushed to the scene with all the help he could muster, the God who gave lavishly to help the relief effort. Truly, if we believe in Matthew's God, the Emmanuel, we must learn to see God in that way. Remember that when Jesus died, the earth shook and the rocks were torn in pieces, while the sky darkened at noon. God the creator will not always save us *from* these dark forces, but he will save us *in* them, being with us in the darkness and promising us, always promising us, that the new creation begun at Easter will one day be complete, and then there will be full healing, full understanding, full reconciliation, full consolation. The thorns and thistles will be replaced by cypress and myrtle. There will be no more sea.

The Gospels pose questions to us at every level, questions about what we have called the atonement as well as questions about what we have called the problem of evil. Dare we stand in front of the cross and admit that all this was done for us? Dare we take all the meanings of the word *God* and allow them to be recentered upon, redefined by, this man, this moment, this death? Dare we take the chaos of the dark forces within ourselves and allow Jesus to rebuke them as he rebuked the wind and waves on the Sea of Galilee? Dare we address the consequences of what Jesus said, that the rulers of the world behave in one way but we must not do it like that? Dare we put atonement theology and political theology together, with the deeply personal message on one side and the utterly practical and political message on the other, and turn away from the way of James and John, the way of calling down fire from heaven on our enemies, to embrace the way of Jesus? Dare we live out the message of God's restorative justice, claiming the victory of the cross not only over the obvious wickedness of the world but over the wickedness of those who fight fire with fire, bringing a solution by creating further problems? Dare we stand at the foot of the cross, feeling the storm clouds darken overhead and the earth tremble beneath our feet, and pray once more for God to finish his new creation, to make the wolf and the lamb lie down together, to bid the mighty waves of the sea be still and depart for good, to establish the new heavens and new earth in which justice and joy will dwell forever?

Evil is still a four-letter word; so, thank God, is love. God grant us grace to be so filled with that love that we may work in our own day with mature, Christian, sober intelligence to address the problem of evil, to implement the victory achieved on the cross, and to be agents, heralds, and living embodiments of that new creation in which the earth will be filled with the glory of God as the waters cover the sea.

7

How the Bible Reads
the Modern World

Among the many interesting differences between America and Britain is that the United States right now seems far more obviously polarized. Britain has its divisions, rivalries, and ongoing tribal suspicions, but in America they seem much sharper and more dangerous. Many Britons watched the 2012 American election results come in, and among the abiding images were the euphoria on the one side and the abject misery on the other. In Britain, people are painfully aware that no politician can be a Messiah. They assume nobody is going to solve all their problems, and they vote for the one they think will make the least mess. Americans, I think, are encouraged by the media and the parties themselves to think that one candidate with the right vision can at last bring the utopia they know they deserve. And to that I want to say, quoting a famous American song, "It ain't necessarily so."

Now, that song tells its own story, again one that sounds strange to British ears. *Porgy and Bess,* the Gershwin musical from which it's taken, is a quintessential American icon. And it is typically in America, not in Britain or most other parts of the world, that people in the twentieth century would feel the need to say to one another, "The things that you're li'ble to read in the Bible—it ain't necessarily so." Christians in Britain have their own debates about what the Bible says, what it means, and what trusting it as in some sense God's word might involve. But they don't have the particular debates found in America. Nor does the Bible get brought up much in Britain's public debates the way it does in America. Putting it crudely, the British never had a Scopes trial, nor is it imaginable that they could have.

The polarizations of American culture, going back at least as far as the Civil War, run deep in American society, and they have created a climate in which the so-called culture wars seem to affect everything else, from income tax to gender issues to gun control. (As a friend of mine from out West says, "In Idaho 'gun control' means you use both hands.") The combinations of issues seem to make sense in America, but they don't make sense to many people elsewhere in the world, and bundling the issues means that Americans don't have to stop and think about each one. It's taken care of once you know which side you're on, a decision that people make for different reasons. Perhaps someone from Britain may be allowed to remind Americans of the larger perspective and suggest ways of engaging critically not so much with the apparently presenting, and polarized, issues, but with the sources of that polarization and what we can do about it. And, as my title suggests, we can examine what place the Bible has in it.

The Bible, after all, has played an important role in American culture, or rather an important range of roles. But those roles have often been determined by the culture itself. I want to first examine what those culture-determined roles might be and then ask whether the Bible might suggest that we approach things differently. This will inevitably be a broad-brush sketch, but I hope it will at least raise issues in a new light and stimulate further questions and actions.

How We Got Here

I love the story of the young William Temple, who became arch-bishop of Canterbury. He once asked his father, "Daddy, why don't the philosophers rule the world?" His wise father answered, "Of course they do, silly—two hundred years after they're dead!" Well, we are now living roughly two hundred and fifty years after the European and American Enlightenment, and though that movement took different forms in different countries, it had emphases that remained constant and are now taken for granted, especially in America. The Declaration of Independence and the Constitution were framed not merely by the contingent need to get rid of British rule but by a philosophical belief that the Enlightenment had discovered how the world should work and was now in a position to get it right. Hence the slogan, on the Great Seal of 1782 and on dollar bills to this day: *Novus ordo seclorum*. A new order of the ages! That, in fact, is a quote from a first-century Roman poet celebrating the new golden age under the emperor Augustus, but it was picked up in the 1780s to highlight the brave new world of American independence.

This Enlightenment project had two features in particular that have had, more in the United States than elsewhere, a huge impact on how people see academic study and knowledge, and how the Bible is read both inside and outside the church. These features have also shaped aspects of the political culture. It all gets bundled together, as I said. And I want to suggest that though these features possess a grain of truth, both are in fact seriously misleading.

The Split-Level World

The first feature is the Enlightenment's embrace of Epicurean-ism. Thomas Jefferson said famously, "I am an Epicurean." Epicurus was a third-century BC philosopher popularized by the first-century poet Lucretius. Lucretius, whose work had been lost for a millennium, was rediscovered in the fifteenth century and became one of the cornerstones of the new European culture. Epicureanism splits the world into two. Whereas the Stoics, being pantheists, reckoned that god and the world were basically the same thing, the Epicureans declared that the gods, if they existed at all, were totally removed from the world and never intervened in its affairs. The natural world was thus free to evolve under its own steam. This was a reaction to the gloomy ancient pagan tales of angry divinities out to get you either here or hereafter. No, said Epicurus: the gods don't care what you do, and you are in any case simply a bunch of atoms that has emerged from random processes and will disintegrate at death. Nothing to be afraid of, then; just find the best way to enjoy yourself quietly here and now.

What happened in the eighteenth century was a toxic combination of Epicureanism and three other things. First, there was theological revolt: once again, people wanted to get away from the old bully in the sky. Second, political revolution: people wanted to be free to choose their own system, away from kings and princes and their claim to divine legitimacy. Third, there were scientific and technological advances. The rise of modern medicine, together with discoveries in biology and elsewhere, was, to be sure, exciting and important. In each case, the pattern was the same: we don't want the stuff that is being handed down to us; we will make the world work in our own way and on our own terms.

You can understand how Epicureanism appeared to fit. But in fact these movements by themselves wouldn't compel you to become an Epicurean. Just because we observe evolution, that doesn't mean there can't be a god who is active within that process as well as beyond and above it. Just because we want to cast off tyranny, that doesn't mean there can't be divine impulses and constraints in democracy. Just because we don't like the prevailing theology, that doesn't mean there isn't a better and indeed more biblical one.

Epicureanism is a choice people make on quite other grounds. Science by itself doesn't force you to be an Epicurean. The God of the Bible and the processes of his creation do not constitute a zero-sum game, as so many in our culture still assume.

In addition, many cultures have developed highly advanced scientific work; think of ancient China, ancient Iran with its remarkable astronomy, or the great Muslim culture of the Middle Ages. But if you want to be an Epicurean for other reasons, like many Westerners in the last two hundred years, then science and technology provide a convenient excuse. It is only an excuse, though, because by observing, say, biological evolution, or indeed cosmic evolution from a putative Big Bang 13.7 billion years ago, you have said nothing whatever about the presence or activity of a god—unless you really thought that divine action and natural development were in direct conflict, so that if you postulated the one you were automatically denying the other.

But Epicureanism took hold in Western culture, especially in America, and was one of the reasons for the official church-state split, which still causes arguments about prayer in schools, about "In God we trust" on banknotes, and so on. And this split is still invoked in terms of the project of knowledge itself, and hence of academic study. It is still widely assumed that science and technology provide real knowledge, and that the world of the arts, of poetry and story and painting and, still more, metaphysics or theology, are merely subjective musings without any purchase on solid reality. On the day I was writing this, an article appeared in the London *Times* in which one of our most successful industrial entrepreneurs, bemoaning the dearth of young scientists coming through our education system, caricatured the choice between arts and sciences by speaking of girls going to college to study—this was his phrase—"French lesbian poetry" when they should be learning nuclear physics or engineering. Several leading academics have already attacked him for the outdated outburst, but it shows the way many people still think.

If that's the view of the arts in general, however bizarre the caricature, it is certainly the view many people have of Christian faith and theology. The latent Epicurean split dictates that such things belong "upstairs," as it were, in the private world of a detached spirituality that has by definition nothing much to do with the real world, the dirty and messy world, whether of politics or of science. A detached spirituality in the present is then matched by an otherworldly hope for the future, a heaven that will have left behind forever the world of space, time, and matter. No wonder the last two hundred years have had such trouble with ideas about new creation, let alone resurrection.

And the churches have regularly gone along for the ride. If the culture dictates that you have to choose between God and the world, the churches will choose God—forgetting that in the Bible God is the creator, or, if they do remember, getting stuck on essentially modernist ideas about what creation might mean, with a god who makes a world that looks as though it's been around for 13 billion years but is actually only six thousand years old. Conservative churches have spoken of miracles, in the sense of a god normally outside suddenly reaching in, doing something dramatic, and then going away again. And so the confusion goes on. The Bible, for those who collude with the world of Enlightenment modernism, becomes simply a book of individual escapist spirituality, an unreliable book at that, full of distortions and prejudices. The Bible is first privatized, then dismembered. And anyone who tries to suggest that this view is wrong is immediately labeled a premodern fundamentalist.

One more thing about the split in modernism that has produced what we call secularism. Studies of the human brain have shown that, broadly speaking and allowing for local variations, the right hemisphere handles things like language, music, art, metaphor, poetry, and indeed religion, while the left hemisphere handles brute facts, numbers, calculations, and so on. Many have pointed out that we seem to be hardwired to do both sets of things, and that human life seems to flourish when they are brought into proper balance. One recent study proposes that the right hemisphere is demonstrably designed to take the lead, so that humans first intuit a big picture that is often only accessible through art or story or metaphor, and then go to work to figure out how it all happens. The right hemisphere, in other words, is supposed to lead the way, while the left one crunches the numbers. And some who have glimpsed this remarkable analysis have suggested that in the post-Enlightenment world this balance has gone wrong, so that left-brain thinking and knowledge have been first privileged and then given sole authority. This is the cultural equivalent of schizophrenia. But these assumptions run deep in today's world, and they have radically conditioned the way we approach everything, including not least the Bible.

What happens to the Bible within this split-level world? On the one hand, as with Thomas Jefferson himself, the Bible is cut up into bits: the bits we can believe in and the bits we can't. Moral teaching, or some of it, is okay; miracles are not. Jesus teaching people how to pray is all right; Jesus announcing God's kingdom on earth as in heaven is not. God providing spiritual consolation and encouragement is fine; God at work in the world is definitely not. And so on. Modern biblical criticism was not, as often imagined, a matter of neutral, objective scholarship. It was driven, from the first, by the desire not to understand but to cut down to size, the size in question being the dominant Enlightenment worldview. The trouble is that even the conservative scholars who have tried to defend the Bible against this kind of attack have regularly done so within the same split-level world, so that those who have defended the miraculous, who have wanted to speak of God's action in the world, have done so in terms of invasion—of a god who is normally outside the processes of the created order reaching in, doing a few tricks, and then going away again. And that picture has very little to do with the God of the Bible. Such would-be defenders of the Bible have, in any case, usually not wanted to get too close to the idea of God becoming king on earth as in heaven, which is the main theme of all four Gospels.

This shows, if proof were needed, that the debates about the Bible rumbling on in our culture have usually taken place *within* the same Epicurean framework, rather than challenging that framework. However, as I will show, the Enlightenment's appeal to history backfires. When we take the historical questions with utmost seriousness, the split-level world is shown up as a philosopher's construct.

The Climax of History?

The second element in the Enlightenment package that our culture has wholeheartedly embraced is the belief that with the eighteenth century and its discoveries, the whole of human history had at last come to its climax. This is where it had all been heading: humankind had now come of age, ushering in a new *saeculum*. This belief is expounded by many thinkers to this day, but more importantly it is assumed by billions of people in the Western world. It is the myth not just of development but of *successful progress*. Feeding on the evolutionism that, quite apart from biological evolution, forms an important part of the Epicurean package, this belief declares that with the rise of modern science and modern Western-style democracy, the world has arrived at its maturity. All that has to be done is to implement more fully the utopian vision we ourselves, in the enlightened West, have glimpsed. We have seen the future, full of hope and prosperity and justice and peace, and if we haven't quite implemented it yet, it must be because not everyone has quite got the point . . . so those of us who have got it are under an obligation to force it upon them.

That was more or less exactly how the Romans thought in the first century, in the time of Jesus. It was more or less how the British thought in the nineteenth century. It is now how much of the

Western world, but especially America, thinks today. Every time a commentator or talk show host says, "In this day and age . . ." or "Now that we live in the twenty-first century . . . ," we ought to know what's coming: the assumption that we're all signed up to a social or even moral evolution, always in the direction of what calls itself freedom, however nebulous and ambiguous that word may be. That assumption leads easily to the supposition that all non-Western nations are really liberal democrat butterflies simply waiting for someone to unzip their present chrysalis. Then, when we help them remove their dictators, we can't understand why they don't at once become like us. The myth doesn't work. Looking more widely, one might point out that a movement of thought which produced the bloodthirsty French Revolution, the Gulag, and Auschwitz—all of which came very much in the name of progress—doesn't exactly have the best track record to claim our trust in giving moral guidance for the twenty-first century. Yet people still speak as if anyone who thinks otherwise is somehow mentally deficient or living in the Stone Age.

All this is important in itself. But my point is the effect this had on the reading of the Bible—again, not only in the wider world but among Christians themselves. Let me put it starkly. *The Bible tells the story of the world as having reached its destiny, its climax, when Jesus of Nazareth came out of the tomb on Easter morning.* The Enlightenment philosophy, however, *tells the story of the world as having reached its destiny, its climax, with the rise of scientific and democratic modernism.* These two stories cannot both be true. World history cannot have two climaxes, two destinies. That is why, for the last two hundred years, people have poured such scorn on the story of Jesus's resurrection. Of course, a dead person being raised to a new sort of bodily life always was extraordinary. But the reason the first Christians believed it wasn't that they didn't know the laws of nature. They believed, on the powerful evidence of eyes and ears and hands as well as hearts, that the God of creation had done something new, though deeply coherent, *in and within* the

natural world, launching his long-intended project of world renewal. It wasn't an arbitrary intervention, either simply to rescue
Jesus or to display God's omnipotence, or whatever. It was all
about *new creation*.

But from the eighteenth century on, people have said that if
you believe in modern science—by which they mean the Epicurean project of scient*ism,* which claims empirical evidence for
its philosophical worldview—then you can't believe in the resurrection. This skepticism has, however, nothing modern about it.
Lucretius, the greatest ancient Epicurean, would have scoffed at
the idea of resurrection. So would Homer or Aeschylus or Plato
or Pliny. The point is that *the resurrection, if it had occurred, would
undermine not only the Enlightenment's vision of a split world but also
the Enlightenment's self-congratulatory dream of world history reaching its
destiny in our own day and our own systems.* That's why the resurrection has been seen in scholarship not as the launching of new
creation but simply as the most bizarre of miracles, then as an
impossible miracle, then as a dangerous ideological claim. You bet
it's dangerous. If it's true, other ideologies are brought to book.

This time the churches both have and haven't gone along for
the ride. Many churches have clung to a belief in the resurrection, though they have seen it simply in terms of Jesus going to
heaven when he died, which was never the point. But, more important, most Western churches have simply forgotten what the
Gospel message is all about, and what the Bible, seen as a whole,
is all about: that this is the story of how the creator God launched
his rescue operation for the whole of creation. As a result, the
great narrative the Bible offers has been shrunk, by generations of
devout preachers and teachers, to the much smaller narrative of
"me and God getting it together," as though the whole thing—
creation, Abraham, Moses, David, the early church, and not least
the Gospels themselves—were simply a gigantic set of apparently
authoritative teachings about how unbelievers come to faith, how
sinners get saved, how people's lives get turned around. Of course,

the Bible includes plenty about all of that, but it includes it within the much larger story of creator and cosmos, covenant God and covenant people—the single narrative that, according to all four Gospels, reaches its climax with Jesus.

And this is why—just to rub the point in a bit in terms of North American culture—the dispensationalist teaching about the end times has loomed so large. Of course the Bible looks ahead to the ultimate future. But it sees that ultimate future as the moment when what happened in and through Jesus will finally be fully implemented. Post-Enlightenment culture, not least Christian culture, by taking the emphasis off the cosmically decisive events of Jesus's death and resurrection, has had to place far too much emphasis on the end point, as though it were completely detached from present events. That's what happens when you tamper with the story—or collude with a different story that has become a cultural assumption.

Through it all, that left-brain/right-brain distinction comes back to haunt us. It translates, in popular discourse, into a distinction between objective and subjective. That is precisely how many academics think and speak and write. It wouldn't surprise me to know that there are science professors who mock all other types of knowledge as though they're simply the fluffy, pretty, inconsequential bits around the edge, while (they say) the physical sciences are the solid, hard, no-nonsense things in the middle. Of course, nobody really lives like that for a single day. Music, laughter, grief, and imagination keep breaking in despite the best efforts of the left brain, just as the right-brain dreamers still have to do the laundry and pay their bills and catch the train to get to work on time. But in public discourse and often enough in public policy, the split remains, with the working assumption that the left brain must lead and the right brain must follow or even be squeezed out entirely.

In my own academic field, it is much easier to get a doctorate in biblical studies if you do a relentlessly left-brain analysis of

a small part of the text, whereas if you attempt a fresh vision of
the big picture, within which it might all make sense, someone
is bound to ask you, in tones that reflect only too accurately the
cultural assumptions that lie behind them, "But where is that in
the text?"—meaning, "Give me one verse that says precisely what
you're saying," whereas the answer often lies not in a single verse
(as if one's interpretation of a great painting could be narrowed
down to one square inch of the canvas!) but in the full sweep of
the chapter, the book, the collection of books in question. I have
argued elsewhere that it is time for a fresh integration of different
modes and methods of study, taking full account of these cultural
assumptions and allowing the texts themselves to offer their own
challenge, their own alternative points of view.

The Bible Addressing Western Culture

But I want now, in the second half of this chapter, to suggest three
ways in which the Bible itself, when read for all its worth, might
challenge the cultural assumptions that have worked their way so
thoroughly into popular consciousness, including popular Christian consciousness. What might happen if we listened afresh to
the message of the whole Bible?

The Bible and the Split-Level World

The first thing to say is that the Bible knows nothing of the split-
level world of Epicurean philosophy. When Paul addressed the
high court in Athens in Acts 17, defending himself against the
charge of introducing "foreign divinities" into the city, he man-
aged skillfully to navigate the competing worldviews of his hear-
ers. He denounced the idols and temples—which were of course
the most visible signs of the local culture—on the grounds that

the High God does not live in such places, since he himself is the creator who gives life and breath to all things, and who is "not far from each one of us" since "in him we live and move and exist." But if Paul is not an Epicurean, he is not a pantheist either, like the Stoics. God and the world are not the same thing. They exist in a fascinating if elusive mutual relationship, with God retaining the initiative and also the right and the responsibility to put the world completely right in the end.

Paul here draws on the ancient wisdom we find in Israel's scriptures. Our modern culture has become so hung up on how to interpret the six days of Genesis 1 that it has forgotten three things about that chapter. First, it has forgotten that the sixfold sequence was a way, in that culture, of describing the construction of a *temple:* the creation was designed to be a place where the living God would dwell. The two spheres of created order, called in the Bible "heaven" and "earth," were not separate, detached from one another, as in Epicureanism ancient and modern. Nor was it the case that heaven was good and earth bad, as in Platonism; nor was earth important and heaven irrelevant, as in secularism. Heaven and earth were the twin halves of the good creation, made to overlap and interlock, so that the God who lived in heaven would also be present, though mysteriously so, here on earth, and the dwellers on earth would always be within arm's length of heaven. The whole world, heaven and earth together, was the diverse-but-united home where God and his creatures would live in harmony.

The second thing people forget about creation, and about the worldview it gives us, is that instead of a man-made temple with the statue of a god inside it, we have a heaven-and-earth world with the image of God within it—in other words, with male and female human beings as the way in which the invisible God becomes present to his world, the sovereign God becomes the caretaker, the "taking-care-person," the steward of his world. This at once removes human beings from any possibility of being mere detached, fly-on-the-wall observers, such as is often dreamed of in the quest for objective science, and they become instead responsible participants in God's plans for his world. This has enormously significant consequences for study of the natural world. We are indeed summoned, as human beings, to become as familiar as we can, each through his or her own gifts, with God's world. But this is never so that we can report back to one another on the curious things we have discovered. It is so that we can worship God the creator and be wise stewards of his world.

The book of Genesis offers us, then, a picture of the world as a temple and of humans as the statue, the image of the god, within that temple. Both of these pictures challenge the split-level world that has been assumed in post-Enlightenment society and culture. But there is a third thing as well. Genesis offers us, right up front, not one creation story but two. Genesis 1 and 2 are not strictly compatible—at least, not if you try to take them as left-brain, rationalistic narratives about what happened. They are not, and are not intended to be, what we would call scientific accounts. Part of the point of there being two of them is to alert the reader—and sadly many readers in the last two centuries have not taken the hint!—to the fact that they are poetic images, narratives replete with metaphor, stories designed to help us grasp with our right brains what creation is *for*. As Jonathan Sacks, until recently the chief rabbi of the United Kingdom, says again and again in his book *The Great Partnership: God, Science and the Search for Meaning,* science takes things apart to see how they work, but religion puts things together to see what they mean. And often the process of putting things together involves that quintessentially right-brain activity of telling and grasping *stories,* not so much to know what happened as what it's all about. Genesis invites us, then, into a world that does indeed have two levels, loosely called "heaven" and "earth," but in which those two levels are not split apart as in Epicureanism but fused together in complex and intricate ways. It invites human beings into a world of knowledge which is not that of the detached observer but that of the involved participant.

Of course, all I have done so far is to glance at Genesis 1 and 2. If we extended the search farther, taking in the soaring poetry of the Psalms or the down-to-earth wisdom of Proverbs, we would find again and again that same overlap of heaven and earth, and the vocation of humans caught in the middle of it all, involved in a drama they often don't understand but called nevertheless to be actors, not spectators; to be subjects themselves, not merely seekers after that elusive something called objective truth.

The Bible as Stories and Story

The Bible tells a multitude of stories, but in its final form it tells an overarching story, a single great narrative, which offers itself as the true story of the world. This is so, incidentally, whether you take the Jewish form of the Bible, what Christians call the Old Testament, or the Christian form, including both testaments. In the Jewish Bible, the story remains incomplete. The great philosopher Ludwig Wittgenstein declared that as it stood, the Jewish Bible was like a body without a head. It's a story that is going somewhere but hasn't arrived there yet. The early Christian writings, however, tell and interpret the story of Jesus quite specifically in such a way as to say, "This is where that original story was always going."

Like all the best stories, the climax and resolution, when they come, are unexpected. Nobody in Second Temple Judaism had predicted anything like the kind of figure Jesus of Nazareth turned out to be, doing the sort of things he did; yet, with hindsight, many understood that this was indeed where the story had been heading all along. That, again, is part of the meaning of Jesus's resurrection. Without that, nobody would have suggested for a minute that he was in fact Israel's Messiah, the world's rescuer and lord.

The theme of Israel's story coming to its climax indicates how the very specific story of one small nation in one small part of history and geography can be thought of as the bearer of a purpose that extends right across space and time. The theme of Israel's election has been largely neglected in the last two hundred years, partly because of residual European anti-Judaism or anti-Semitism and partly because of the boost the Enlightenment gave to such subcurrents. The Enlightenment was a totalizing movement, claiming a kind of universal knowledge, and the very particular narrative of Israel was felt as an affront; indeed, in the developing German Enlightenment of Hegel and others, Judaism was thought of as "the wrong kind of religion"—an idea that of course came to horrible fruition in the twentieth century.

But in scripture the narrative of Israel recounts how the creator God develops his plan to rescue creation. As the Psalms and the Prophets regularly point out, when that plan reaches its height, the warring nations of the world will be brought to heel by God's anointed king, whose kingdom will stretch from one sea to the other. That is the story the New Testament sees as having reached its surprising climax in Jesus. Surprising not only because it was unexpected but because it was different in *character;* the New Testament's narrative claim, confronting all imperial narratives ancient and modern, is not about someone conquering the world by the normal means, but about the crucified and risen Jesus starting to rule the world by the means and methods of his self-giving love. And that gives its character to the whole project of Christian knowing, a fully human mode of knowing that includes but transcends the false either/or of post-Enlightenment epistemologies.

In particular, the biblical narrative, in addition to offering a story that makes sense at so many levels, reduces all other metanarratives to humility. In the ancient world, the biggest story around was that of the Roman Empire, an eight-hundred-year narrative of how Rome had grown from small beginnings to be the great imperial power tasked by the gods with bringing justice and peace to the world. The New Testament again and again lines up the story of Israel, with its climax in Jesus, against this imperial narrative, claiming to be the reality of which Rome's self-serving story is the parody. And it does exactly the same to the self-serving stories of the Enlightenment, with all their fallout in terms of that split-level world and its effects on everything from the academy to politics. The Enlightenment's story is similarly a parody of the true one, just as Marx offered the world a parody of the biblical vision of society and Freud a parody of the biblical vision of a redeemed human interiority.

Paying attention to the biblical story, learning to live within it on the assumption that it is the true drama of the world, doesn't mean that one must retreat to some desert island where other ideologies have no purchase. The story of Israel, and of Jesus, is precisely the story of God's people making their way in contested territory, by humble and cheerful witness, sometimes through suffering or even martyrdom, frequently living as a sign of contradiction, not because we are awkward or unreasonable folk who reject on principle everything the rest of the world stands for, but because we seek a genuine humanity that, even in fleeting glimpses, is self-authenticating. And just to be very practical for a moment, part of the point of reading the four Gospels regularly and thoroughly, together and individually, is so we can learn to live within that larger story.

The Bible and Human Knowing

This brings me to the heart of what I want to say here: that in the Bible, when we let it *be* itself, we find a mode of knowing that is neither the brightly lit supposed objectivity of post-Enlightenment scientism nor the fuzzy and indistinct supposed subjectivism that is its opposite. The Bible confronts the split in our world and our knowing that has so bedeviled the modernist project. In the Bible we find a vocation to human knowing that is always relational, always responsible, always fully attentive to the thing or person that is known and yet always bringing to it the larger world of narrative, imagination, metaphor, and art that enables us to know things more fully than merely as a list of facts or a string of formulas. People often wrongly assume that Christians claim to know religious truth in an absolute way, while secular scientists claim a different sort of absolute knowledge, a completely objective account of the world. I want to propose a quite different way of looking at the whole thing, in the form of a different kind of worldview and a different account of knowledge. I hope this will both challenge and encourage you.

Three things about the way this works out. First, the knowing subject is made in the image of the creator, and thus the subject is called to *reflect* the creator's wisdom and care into the world, and to reflect the praises of creation back to the creator. Within this model, knowing can never mean simply amassing knowledge *about* the world, about different types of trees, or the pattern of DNA, or the dark matter beyond the stars. All that is important, but the knowing subject has a *vocation* in relation to the known. Knowing *about* the world is supposed to be part of the work of bringing the creator's wise ordering *into* the world, thus enabling the world and its various parts to flourish, to be more gloriously what they truly are, and to bring praise to their creator. This is part of what it means, in biblical teaching, to be made in God's image. To repeat Jonathan Sacks's maxim: science takes things apart to see how they work, while religion puts them together to see what they mean. The two need each other, just as the two halves of the brain need to work in proper balance.

Second, the relation of the subject to the material being studied is significantly different. The word *knowledge* varies considerably in accordance with the nature of the thing known. Here I want to stress that the scientific knowing, in which one conducts experiments that are in principle repeatable (to be sure one is not deceived or muddled) is on a continuum with several other kinds of knowing. Within modernist and especially Epicurean thought, some have been tempted to privilege scientific or empirical knowing as the only real kind. Other kinds of knowing—for instance, the sort of knowledge a philosopher or literary critic might claim—is then regarded as somehow second-rate. But the knowing that goes with wisdom in the biblical sense sees the object of study not as an isolated entity to be manipulated or exploited but as part of a much larger world of interlocking connections and mutual relationships.

Third, the knower never knows in isolation. All serious academic study takes place in the context of a community of knowers. And here is the point. True wisdom is both bold and humble. It is never afraid to say what it thinks it has seen, but will always covet other angles of vision.

Where has this taken us? It has taken us to the point where we can see an integrated mode of knowing. All knowledge, suggests the biblical wisdom teaching, involves human beings in a much more complex series of relationships than simply that of the detached observer obtaining a supposed "view from nowhere."

There is an obvious biblical term for all this, and it is *love*. Our word *love* has tried to do so many jobs within the English-speaking world that it is hard to get this particular meaning into focus. But we must try. One of the primary things about love is that it strongly and radically affirms the person or thing that is loved. It doesn't try to manipulate him, her, or it, or to pull it out of shape; it desires the best for its object and works to bring that best about. But, at the same time, genuine love, whether for a person or an object, a tree, a star, a mountain, or a piece of literature, can never be objective in the sense of offering a detached, fly-on-the-wall perspective. Love draws together what the Enlightenment split apart at the level of knowing, just as—and for the same reasons—in my own field it tried to split apart the Jesus of history and the Christ of faith. It can't be done. Love is indivisible, and Jesus is indivisible, and the two rather obviously go together. And if you think that sounds like a pietistic short-circuiting of the whole thing, so that a leap of faith toward Jesus makes everything else irrelevant, that simply shows how deeply the split world has still got hold of us.

It is, in fact, Jesus who is the subject of it all. Within the larger narrative of the Bible, as Saint Paul says, all the promises of God find their yes in him; among those promises are promises about knowledge. In him are hidden, Paul says, all the treasures of wisdom and knowledge—both the treasures that wisdom and knowledge perceive and celebrate and the treasures that we call wisdom and knowledge. Or, as Saint John puts it in his breathtaking opening, "In the beginning was the Word"—the speech, the self-expression, the creative power and personal presence of the transcendent creator. And the Word became flesh and pitched his tent in our midst. All cultures, ancient and modern, have found this challenging. It isn't surprising that ours has done so in its own way. The task of those who follow Jesus is to go on struggling against all that threatens to squash this biblical vision into the alien boxes of other worldviews; to have our big picture, our right-brain vision, refreshed and renewed by the sweeping story and powerful imagery of the Bible, and so to get on with our left-brain work, whatever it may be, in the confidence that because Jesus is lord of the world, all truth belongs to him.

We won't get it all right. Our task is not to be triumphant experts on everything. That too would be to capitulate to another version of the modernist dream. Our task is to be faithful to the vision, humble but cheerful in our pursuit of it, and confident in God the creator and life giver—confident that, as Paul says, in the Lord our labor will not be in vain.

8

Idolatry 2.0

BEFORE I DIVE in, allow me to set the scene. We in the
Western world are, for better or worse, children of the
Enlightenment. And the main idea of the Enlightenment was rooted in Epicureanism. You could sum it all up like
this: the gods are a long way away and they don't bother about us,
so relax and enjoy your life. There are three things to note about
ancient and modern Epicureanism and how it has affected the
way we think now.

First, Epicurus and the first-century BC poet Lucretius were
reacting strongly, but not against ancient Judaism, which they
didn't know, and not of course against Christianity, which hadn't
begun then. They were reacting against certain types of ancient
paganism that frightened people by suggesting that there were
gods all over the place, that they were out to get you, and that if
they didn't make life miserable for you here and now they might
well do so after you died. You then get the same reaction in the
fifteenth century, when the medieval church had borrowed a lot
of those ancient pagan ideas about unpredictable divine anger to
frighten people into believing or behaving, so the rediscovery
of Lucretius, and thus of Epicureanism, came as welcome news.

When Thomas Jefferson declared in the eighteenth century that he was an Epicurean, he and his Enlightenment colleagues were consciously reacting against a kind of Christian teaching that threatened people with an angry god both here and hereafter.

My second point is that Epicureanism wasn't the foundation of modern science, but it was the foundation of what we might call modern *scientism*. By splitting off our world from any possibility of divine intervention or encounter, Epicurus and his successors gave to the world a kind of autonomy, an independence. This independence is the ancestor of the Enlightenment's claim to scientific autonomy.

It is also the root of political autonomy, which is the cousin of scientific autonomy—and the parallel development of the two in the Western world ought to give us food for thought. Just as Epicureanism gets rid of divine interference and lets the natural world evolve in its own way, so you can get rid of the divine right of kings and let democracy develop in its own way. That belief has driven an assumption that when countries elsewhere in the world get rid of their dictators, they will naturally want to become Western-style democracies. When that doesn't happen, as it hasn't, we have no other narrative to help us understand what's going on. In fact, letting systems do their own thing, as Marx saw clearly, might well mean that they boil over into revolution, like some elements in the natural world. An Epicurean vision of politics needs to allow for the equivalent of volcanoes.

This brings me to my third point before we move on to modern culture: this confluence of ideas has given birth to what we think of as *secularism*. Secularism is a complex phenomenon, but it has become a dominant motif in Western culture, particularly in the United States. Despite—or perhaps because of—the continuing and often strident religiosity of American culture, there has been increasing pressure to banish talk of a god from public life and to conduct everything from scientific research to politics and even marriage on the assumption that the world is and means what it is and means without reference to anything beyond its visible, and in principle scientifically measurable, self.

I have put this essay's question into triple context because it seems vital if we are to understand where we have come from and not accept the sacred/secular divide or the religious/nonreligious divide as simply part of some unalterable given cultural landscape. It is no such thing. Ironically, the current cultural landscape is partly comprehensible as one more evolution in the complex history of the modern Western world. But it has solidified itself, politically as well as scientifically, through the remarkable claim made by the founders of the United States in the late eighteenth century, who really did believe that they were seeing the birth of a new *saeculum,* like the Roman poet Virgil said in the time of Augustus two thousand years before. This was to be the golden age.

That claim, hiding powerfully just under the surface of so many cultural assumptions, particularly in America, means that any attempt to challenge the perceived rule of secularism is seen ipso facto as a challenge to the great modern order that has brought us so many obvious blessings, not least in the medical sphere.

All this leads me to the main section, in which I want to suggest that the assumed standoff between what we call religion and what we call the secular world, and the cultures that have grown up around this standoff, are radically misconceived, and that there are other ways of looking at the whole thing that would be more accurate in description, more helpful in enabling us to find our way forward, and more Christian in their conformity to the interesting and often forgotten message about Jesus.

What Happened to the Gods?

One of the things we learn early in science is that nature abhors a vacuum. You can create a vacuum, and you can sustain it given the right technology, but atmospheric pressure always threatens to break back in, sometimes causing an explosion. Well, something similar is true in philosophies and worldviews. They abhor a vacuum. You can push God, or the gods, upstairs out of sight, like an elderly embarrassing relative. But history shows again and again that other gods quietly sneak in to take their place.

These other gods are not strangers. The ancient world knew them well. Just to name the three most obvious: there are Mars, the god of war, Mammon, the god of money, and Aphrodite, the goddess of erotic love. One of the fascinating things about modern Western ideas has been the work of the "masters of suspicion," Nietzsche, Marx, and Freud, claiming to reveal the motives that lie hidden beneath the outwardly smooth and comprehensible surface of the modern world. It is all about power, declared Nietzsche. Everything comes down to money, said Marx. It's all about sex, said Freud. In each case these were seen as forces or drives that were there whether we liked it or not; we might imagine we are free to choose, but in fact we are the blind servants of these impulses.

Take them in reverse order. It's hard to imagine now the way things were in the 1950s, when I was a child. There was more or less no pornography. The great majority of married couples stayed married. When I was at school, I knew precisely one boy who came from a broken home. No doubt a great deal of what was seen as illicit sexual activity went on below the radar, but a broadly Judeo-Christian moral stance was assumed in society—which meant, importantly for the story I'm telling, that most people felt at least some pressure to resist impulses that, left to themselves, would move in a very different direction. But when Freud became popular, filtering down into mainstream culture through novels and plays, people began to speak of the erotic impulse, often called "the life force," just as they might before have spoken of a divine command. One should not resist. It would be hypocritical and wrong. I don't think people now speak reverently about the life force in the way they did; it's just assumed.

The late Christopher Hitchens, another high priest of contemporary atheism, said that one should never pass up an opportunity to appear on television or to have sex. The goddess Aphrodite, even if unnamed, is served by millions.

Today's Western world hardly needs reminding about the place of Mammon, the worship of money, in our society. Britain, or rather London, has prided itself on being the financial capital of the world, and the major financial scandals and banking crises that have shaken our system over the last decade have done nothing to damage our faith in this ancient and yet very modern god. We still assume that though something has gone wrong, the only thing to do is to shore up the system and get it going again—despite the gross inequities, the countries still suffering from unpayable debt, the rising tide of poverty even in our affluent Western world, and so on. Perhaps it wouldn't be straining the point to say that many students now hope, rightly or wrongly, that a degree will be a passport to a good job and a good salary, and that is justification enough. You can recognize the worship of Mammon precisely at the point when someone asks you to do a job for which you will be paid considerably less than you are at present. What would you say?

The same is true in relation to Mars and the whole world of power, of force, ultimately of arms and violence and war. Western culture assumes that the answer to the world's problems is to drop bombs on someone. I said in 2002, and I have said ever since, that unleashing our modernist military machine on the putative sources of terror in the Middle East was just going to make everything much, much worse, and I have seen nothing in the last decade to make me think I was wrong. But our English-speaking culture, brought up on tales of success in two world wars, has pushed itself into the position of being the world's policeman, and however many body bags are brought home, we still assume that's the way the world ought to work. Experiments like Desmond Tutu's Commission for Truth and Reconciliation in South Africa are admired but not imitated.

In each of these three cases—Aphrodite, Mammon, and Mars—these ancient and well-known gods have not gone away, have not been banished upstairs, but are present and powerful— all the more so for being unrecognized. In what sense are they divine? The ancients would have no trouble answering that. First, those who worship gods become like them; their characters are formed as they imitate the object of worship and imbibe its inner essence. Second, worshipping them demands sacrifices, and those sacrifices are often human. You hardly need me to spell out the point. How many million children, born or indeed unborn, have been sacrificed on the altar of Aphrodite, denied a secure up- bringing because the demands of erotic desire keep one or both parents on the move? How many million lives have been blighted by money, whether by not having it or, worse, by having too much of it? (And if you think you can't have too much of it, that just shows how deeply Mammon worship has soaked into us.) And how many are being torn apart, as we speak, by the incessant demands of power, violence, and war? Now, please note: I am not saying sex is evil. I am not saying money is bad in itself. I am not even saying that there is never a place for force in defending the weak against violent evil or unjust tyranny. I am neither a killjoy, a Marxist, nor a pacifist. My point is that our society, claiming to have got rid of God upstairs so that we can live our own lives the way we want, corporately and individually, has in fact fallen back into the clutches of forces and energies that are bigger than ourselves, more powerful than the sum total of people who give them allegiance—forces we might as well recognize as gods.

Those are just the obvious ones. There are other ancient gods still alive and active: the gods of blood and soil, of racial and ter- ritorial identities and claims. Most of us are now embarrassed to think of them, but they rear their heads even in our avowedly secular society. And behind them is a force that we invoke at every turn, a force assumed particularly in the media: the force of progress. "Now that we live in the modern world . . . ," the speaker begins, as the segue to some argument, usually in favor of a secularist proposal. "Now that we live in the twenty-first century . . . ," people say, as though a change in the calendar meant an automatic updating of moral systems and assumptions. A moment's thought shows how ridiculous this is, but people still say it, and what's worse, they think it.

A further moment's thought shows how this integrates with the three major divinities. Western progress means that we can now send unarmed drones to kill people in Pakistan; what would we think if another power did that to us? Medical progress enables us to have sex with more people with less risk—up to a point—so if risk is the only thing stopping us, what's the problem? And electronic systems enable us to gamble with zillions of dollars of someone else's money, so why not? So our brave new secularist world lurches to and fro in obedience to impulses that an earlier age would have recognized as divine but which we, in our late-modern Epicureanism, have not named as such and so have not challenged.

Now, you may say, what has all that got to do with religion? Many in the West still think of religion as saying prayers, going to church, perhaps reading the Bible or some other sacred text, and not least, living in the hope of going to heaven, however vaguely that is expressed. But this is a thoroughly modern definition of religion, which wouldn't have been recognized by anyone in the ancient world or indeed by many today outside the Western world. From ancient times to the present, religion has had to do with the wider assumed dimensions of ordinary culture, whether they be marriage or music, politics or city life, wine or war. It is part of our dilemma that we have separated what we call religion from what we call politics, in exactly the same way as Epicureanism separated the gods from the world—or thought it did. In fact, it was an unstable philosophy in the ancient world, better suited to the small number of idle rich who could afford to settle down in comfort, and it has been exactly the same in our own world.

Epicureanism is a philosophy for the elite, or those who aspire to be the elite.

Perhaps the convulsions we have gone through—the disasters that come from worshipping Mars, Mammon, and Aphrodite—are signs that the theological vacuum caused by separating god from the world is at last imploding. But do we know what to do under such circumstances? Have we got a road map to help us navigate such dangerous and complex territory? This brings me to my third and final point. I want to suggest that we live in a much more thoroughly integrated world than our culture has recognized, and that there are modes of knowing and being within that world which we can explore and which challenge the unhelpful polarizations of secularism.

Believing and Knowing Within an Integrated World

Today, many people, especially academics, assume that real intellectual work takes place in the objective world of the hard sciences, and that the more you move in the direction of the so-called arts, especially things like metaphysics and theology, the more you are simply talking nonsense about nothing. This is a function of Epicurean assumptions, not of the hard sciences themselves; many periods and cultures have developed sophisticated scientific work without assuming that you had to split that off from other kinds of knowledge.

Nevertheless, many leading scientists today were brought up on the split-world viewpoint. Some have even, with unintended irony, made it an article of faith that one should not allow articles of faith into the classroom or laboratory. And some Christians have gone along for the ride. In the late 1990s, I spoke about Christian ecological work in a think tank in Washington, D.C., and was told afterward that I had lost the Christian half of my

audience when I mentioned scientific evidence for global warming; the very mention of science apparently sent a message that I
was colluding with atheism. That is to make the mistake I mentioned before, of confusing science with scientism, of placing the
proper and wise investigation of the natural world within the
worldview of Epicureanism, which itself is unproved and indeed
unprovable.

So what's the alternative? Here, perhaps to the surprise of
some, the Christian worldview has a great deal to offer, when
you trace it back to its beginnings in ancient Israel, then to Jesus
and the writings of the first two or three Christian centuries. The
category that emerges again and again in the scriptures and the
great teachers of the faith is wisdom, *sophia* in Greek, *Chokma* in
Hebrew. Wisdom is a strange, evocative figure, sometimes per-
sonified as a lady inviting people to a feast, teaching people how
to navigate through this wonderful but also dangerous world.
Wisdom in the scriptures was possessed by the master builder
who constructed the beautiful tabernacle. It's what Joseph was
seen to possess when he advised Pharaoh on how to cope with
ecological and economic disaster. It's what Daniel had when he
read the writing on the wall at Belshazzar's feast. In our culture
we often split apart skills like architecture and engineering from
those like philosophy or political insight. The figure of wisdom
holds them together. Wisdom is what you need, according to
scripture, to become genuinely, fully human. And genuine, fully
rounded humanity is what our culture, with its pretense of religion and its variety of unnamed but powerful gods, has been
remarkably short of.

There are three steps to biblical wisdom. They are all summa-
rized in a much-repeated shorthand that may sound forbidding
to us: "The fear of the Lord," say the writers, "is the beginning
of wisdom." Our problem with that is that we've forgotten who
"the Lord" is, mistaking him for a distant, faceless bureaucrat.
Consequently, we've forgotten that fear does not mean a cringing,

nervous, slavish state, but rather the reverence and awe properly due the creator and sustainer of all things. In other words, just as the concept of faith changes with its object—if you put your faith in a being rather like an amorphous and characterless gas, it will be a very different thing from a faith placed in the father of Jesus Christ—so fear changes depending on whether the god you imagine is a celestial bully or the God of faithful love. My point is this: in the biblical traditions and the later traditions that reflect them, the human subject is imagined not as a lonely, isolated individual who can work things out for himself (it was always a "he"), but as someone who is located in relation to three other coordinates.

First, the knowing subject is made in the image of the creator, meaning that the subject is called to *reflect* the creator's wisdom and care into the world and to reflect the praises of creation back to the creator.

Second, the knowing that goes with wisdom in the biblical sense sees the object of study not as an isolated entity to be manipulated or exploited but as part of a much larger world of interlocking connections and mutual relationships.

Third, the knower never knows in isolation. True wisdom is both bold and humble. It is never afraid to say what it thinks it has seen but always covets other angles of vision.

Now we can see an integrated mode of knowing. This integrated mode of knowing is quite different from the pure objectivism dreamed of in scientism, as well as the supposedly pure subjectivism of which the Epicurean scientist accuses his or her neighbors in the arts faculties, especially theology and similar subjects. All knowledge, suggests the biblical wisdom teaching, involves human beings in much more complex relationships than that of the detached observer obtaining a supposed view from nowhere. The scientist studies that which can be repeated; the historian, that which cannot be repeated; but in many other respects the sequence of knowing is the same, from observation to hypothesis to further testing, all suffused with wonder at the rich complexity of life in another time and place. And the philosopher or metaphysician is doing the same, mutatis mutandis, with the larger question of how it all makes sense. Once again the whole process requires humility, patience, an unwillingness to grab at the material or foist one's own ideas or personality onto it.

Humility, in particular, because, as Saint Paul reminds us, all our knowledge at the moment is partial. "We know in part," or, as he says in sharper mode, if someone thinks they really know something, they don't yet know as deeply as they ought to. And that is the point where he turns the whole thing around. The desire for knowledge is a deep and proper human longing, but ultimately it is a subset of something larger, deeper, and stranger. Ultimately, the different modes of knowing suggested by wisdom boil down to what one might call love.

Love is a mode of knowing all right, but love transcends the objective/subjective distinction that has been inscribed in so much of our thinking in the Western world. In love, the main thing is to admire, respect, celebrate, and take delight in the object of that love, to let it be itself, to *want* it to be gloriously and freely itself, not to snatch it or control it or squash it into another shape. That's not love, but lust. Is love then the pursuit of objectivity? Of course not. Precisely at the moment when love is celebrating the radical otherness of the beloved—whether it be a star twenty million light-years away or a human being twenty millimeters away—love enters into a relationship with the beloved, a relationship that is defined by the nature of the beloved, but in which admiration is mixed with curiosity, a desire to discover more, respect with a longing for intimacy, a desire to know and (so far as is possible) to be known.

In the last two hundred years of Western epistemology, we have seen a splitting of love. It is, of course, possible to divide things up, as has been done, so that the various types of knowledge appear so different as to be almost if not entirely incompatible.

That enables the kind of knowledge we might acquire in the hard sciences to rush ahead unencumbered, producing not only penicillin (which was discovered by accident) but also gas chambers (which were made deliberately). It makes great strides, but not always in the right direction. We have a thousand machines for making war but none for making peace. We have computers and iPhone apps that can make millions out of a tiny change in exchange rates, but none that can rescue the poorest countries from their plight. We know how to make Internet pornography, but not how to repair marriages. The very objectivity or neutrality of scientific knowledge as commonly conceived has played into the hands of the gods we secretly worship.

We have run the risk for too long of taking apart our entire world to see how it works, in order that we may make it work to our short-term advantage. Perhaps it is time to allow other perspectives to come into the frame, since the meaning we have made of our dismembered world has so obviously reflected the gods we secretly worship rather than the God in whose image we are made.

For the Christian, of course, the central claim is that in Jesus, the Jewish Messiah, wisdom became human, became a person, and went about bringing precisely that reintegration for which the world had longed, taking upon himself the disintegration and selfish hatred of the world and rising again to launch the work of new creation. Modern science, insofar as I understand its origins, began with people who understood their task in terms of exploring the natural world with respect and delight as part of the mysterious creation of a wise creator. The Epicurean split world of the Enlightenment encouraged a split world of knowledge, a two-track culture in which each side became increasingly opaque, even repellent, to the other. In the United States, this has sometimes been linked to other aspects of the culture wars, greatly to the detriment of all parties.

I believe it is time to work for a fresh integration, and I suggest to you that the figure of wisdom incarnate, Jesus himself, is the place to begin if we are to discover what that might mean, not only for ourselves but for the wider world, which knows only too well what a divided existence is like and longs for the day when things will be put back together and make the sense they were supposed to do. The great claim made by the early Christians was not so much that in and through Jesus they had a new secret knowledge; some people tried to go that route but, as Saint Paul said, it was simply a clever way of boosting their own pride. The great claim was that in Jesus they had discovered a larger reality within which their partial knowledge made sense. That larger reality was what they called *agápē,* love: the love of the creator God for them and the love they then found in themselves for the creator, for his world, and for one another. Authentic Christian faith is not to be played off against scientific knowledge. Rightly understood, faith and knowledge are offshoots of a larger reality, whose name is love.

Jesus has been studied historically from every possible angle. Some have despaired of ever understanding him; some have questioned whether he even existed, though no serious historian would make that mistake. As a historian, I can say with confidence not only that he really did live in first-century Palestine; but that he really did tell his surprised contemporaries that God was now taking charge of the world in a new way, which he himself was modeling; and that he went to his death believing that this was how God's powerful love would overcome the power of evil and launch his new creation. Again as a historian, I have to say that without Jesus's resurrection I cannot explain why his disciples would have taken this claim seriously after his death.

Of course, this poses challenges for us at the level of worldview, as it did for Jesus's contemporaries. They knew as well as we do that dead people stay dead. But Jesus's first followers believed not only that he was truly alive again but that he had, as it were, gone through death and out the other side, leading the way into a new mode of being in which the power of love would defeat the love of power, in which creation and beauty would win out over death and decay, and in which God would become present to people of every shape and type, offering healing and reconciliation, a new start, a new life, a new way of life. Sometimes, in science as in history, the great leap forward to a fresh hypothesis happens when you put a new element in the middle of the picture and discover that all the jumbled pieces come together in a new and coherent way. That's what it's like when you put Jesus in the middle.

Because it's Jesus, it doesn't all happen at once. He wants us to grow up and take responsibility, to think it through, to be learners, disciples, not just mute followers. But because it's Jesus, he offers and provides the strength and courage that enable us to believe, to learn, and to join in his project of healing and hope. And love.

9

Our Politics Are Too Small

B ACK IN THE early 1980s, I would never have suspected
that I would now be passionately interested in the question of God in public. At that time I had begun to realize, with some alarm, that the work I was doing on Jesus in
historical context might force me to take political questions more
seriously. By the time that work matured into *Jesus and the Victory
of God,* I had come not only to accept the political implication as
inevitable but to be excited about a new attempted integration
of what had previously seemed like two entirely separate worlds.
Around that time I began to discern the political dimensions of
Paul's letters, a matter of ongoing controversy. And during the
last decade of the twentieth century and the first decade of the
twenty-first, I found myself increasingly involved in public life
as well as pastoral ministry, and so have been forced to reflect
further on faith and politics not merely as armchair theory but
as practical challenge. What I want to say here is thus a kind of
report on work in progress.

So much by way of autobiographical introduction. Like Paul in
2 Corinthians, I hope you will bear with this small bit of foolish-
ness. And while I'm about it, let me gently point out that the

political spectrum in the United Kingdom, and indeed in Europe, is quite different from the spectrum in the United States. In Britain, issues are bundled up in different ways than in America. What's more, over the last forty years, those in the United Kingdom who have tried to integrate faith and public life have mostly been on the left of the spectrum, while those who have done the same in the United States have tended to be on the right. The subtitle of Jim Wallis's book *God's Politics* is *Why the Right Gets It Wrong and the Left Doesn't Get It;* had he been writing about the British scene, he might have had to put that the other way round. That is why the *Economist,* in a special supplement on faith and politics, kept insisting plaintively that the only answer to our ills is resolute separation of church and state. No doubt the editorial team of that business-oriented magazine was worried that once God gets out of the box there may be some serious questions to face.

Both countries, however, have been the victims of serious terror attacks and responded together in going to war in Afghanistan and Iraq. And throughout the crises and debates of the years since 2001, it has become embarrassingly clear that the question of God in public is one we should have been addressing at all levels for a long time. In 1987, the British theologian John Bowker published *Licensed Insanities,* which pointed out that all the world's major trouble spots had an irreducible religious component and that the reason none of our politicians could figure out what to do was that none of them had studied religion (let alone the Bible) in college. That prophecy went unheeded. And the reason for that collective deafness brings us to the first major point I want to make.

God into Public Won't Go?

The reason nobody bothered to take people like Bowker seriously was that for two hundred years Western society has assumed, more or less, that God doesn't belong in public life. The split world of the Enlightenment, whose apparently sensible justification was to avoid the appalling wars of religion, had a not so benign motive: keep God out of public life, and then we can run the world to our own advantage. As Deism gave way to implicit and then explicit public atheism, Western democracy justified its existence with the slogan *vox populi vox Dei,* thus giving an appearance of public theology while in fact imprisoning God in the ballot box and so neutralizing any possible political critique from the Christian gospel. Seen from this perspective, the massive achievement of William Wilberforce in the first third of the nineteenth century was to keep up an explicitly Bible-based political critique in a world where such a thing was increasingly seen as a philosophical and cultural faux pas.

But the disjunction between God and the public world has increasingly worked its way into the assumed worldview of the Western world, seen not least in the increasingly shrill rhetoric of secularists who, still clinging to their now clearly outdated thesis, really believe that religion should have withered and died long ago, and now turn on it with rage and incomprehension for its presumption in not only still existing but presuming to have something to say in public. It is fascinating to watch the way Dawkins, Hitchens, and others have repeated, with increasing moral indignation, the rhetoric of Voltaire at the end of the eighteenth century and Nietzsche at the end of the nineteenth. They represent the high-water mark of late modernity, and their increased outrage is explained by the alarming rise of fundamentalism in various guises on many different continents. Indeed, I want to suggest here that the Bible enables us to navigate a path of wisdom not just halfway between secularism and fundamentalism but on a trajectory that shows up those ugly brothers as simply missing the point, representing two opposing wings of a now thoroughly discredited worldview.

Because discredited it has been. It may not be necessary any longer to argue the postmodern point that the large narratives of modernity have run out of steam. September 11, 2001, was a defining postmodern moment, with two grand narratives colliding and exploding, and the response to that event in the form of more high modernist warfare is a stunning example of missing the point. I have been accused, not least by some high-ranking persons in the White House and elsewhere, of myself missing the point on this one, but I'm afraid I am impenitent; and my lack of penitence has nothing to do, as my detractors have absurdly suggested, with going soft on terror or failing to take account of Romans 13 with its legitimation of the rights of ruling governments. I have become used to getting plaintive e-mails saying, "We like what you write about Jesus and the resurrection; we are fascinated by what you say on Paul; but why are you so critical of our president?" And my answer is, "If you actually read what I say about Jesus and the kingdom, and you understand what Paul was really on about, you'll have to take the questions of God in public seriously in a whole new way. To say that I was confusing spiritual issues with political issues is simply to restate the old Enlightenment dichotomy at a moment when it is disastrously inappropriate as well as misleading." But more of that shortly.

Meanwhile the major ethical and public/political issues of our day rumble on—global debt, the ecological crisis, the new poverty in our own glossy Western society, issues of gender and sex, stem cell research, euthanasia, and not least the complex questions of the Middle East—and as long as the debates are carried out in terms of fundamentalism and secularism, they will never be anything other than a shouting match. At this point, ironically, the Enlightenment dream has begun to eat its own tail, as its greatest strength—the emphasis on reason as the means to peaceful coexistence—has been undermined by its greatest weakness, the dualistic division between God and the public world, with human public discourse collapsing into spin and emotivism.

Perhaps the unintended consequence of the postmodern revolution is to reveal that if reason is to do what it says on the label, we may after all need to reckon with God in public. I therefore want to advance a proposal that might help us navigate a path of biblical wisdom as we move forward into the dangerous and uncertain future of post-postmodernity (life after life after modernity, if you like). Tomorrow's world urgently needs to find a way forward that is neither that of secularism nor that of fundamentalism, nor of mere deconstruction, which declares that both are wrong but has nothing to put in their place. You won't be surprised that I believe such a way can be found by returning to the foundation documents of the Christian faith, in particular the four Gospels.

The Public Face of God: Gospel and Kingdom Then and Now

In *The Great Awakening,* Jim Wallis describes how, as a young man growing up in an evangelical church, he never heard a sermon on the Sermon on the Mount. That telling personal observation reflects a phenomenon about which I have been increasingly concerned: that much evangelical Christianity on both sides of the Atlantic has based itself on the Epistles rather than the Gospels, though often misunderstanding the Epistles themselves. Indeed, in this respect evangelicalism has simply mirrored a much larger problem: the entire Western church, Catholic and Protestant, evangelical and liberal, charismatic and social activist, has not actually known what the Gospels are there for. To a considerable degree, the entire corpus of Gospel scholarship since the Reformation and the Enlightenment, reaching a low point in attempts to see the canonical Gospels as late and socially conformist and the gnostic tracts as earlier and more exciting, is in danger of becoming a huge and complex exercise in missing the point.

Matthew, Mark, Luke, and John are all in their various ways about God in public, about the kingdom of God coming on earth as in heaven through the public career *and* the death and resurrection of Jesus. The massive concentration on source and form criticism, the industrial-scale development of criteria for authenticity (or, more often, inauthenticity), and the extraordinary inverted snobbery of preferring gnostic sayings and sources to the canonical documents all stem from, and in turn reinforce, the determination of the Western world and church to make sure that the four Gospels will not be able to say what they want to say, but will be patronized, muzzled, dismembered, and eventually eliminated altogether as a force to be reckoned with.

The central message of all four canonical Gospels—in their very different ways—is that the creator God, Israel's God, is at last reclaiming the whole world as his own, in and through Jesus of Nazareth. That, to offer a dangerously broad generalization, is the message of the kingdom of God, which is Jesus's answer to the question "What would it look like if God were running this show?" And at once, in the twenty-first century as in the first, we are precipitated into the vital question "Which God are we talking about, anyway?" It is quite clear, whether you read Christopher Hitchens or Friedrich Nietzsche, that the image of God running the world to which they are reacting involves a celestial tyrant imposing his will on an unwilling world and unwilling human beings, cramping their style, squashing their individuality and their very humanness, requiring them to conform to arbitrary and hurtful laws and threatening them with dire consequences if they resist. This narrative (which contains a fair amount of secularist projection) serves the Enlightenment's Deist agenda, following centuries of religious wars, as well as the power interests of those who would move God to a remote heaven in order that they can carry on exploiting the world.

But the whole point of the four canonical Gospels is that the coming of God's kingdom on earth as in heaven is not to impose an alien and dehumanizing tyranny but rather to confront alien and dehumanizing tyrannies with the news of a God—the God recognized in Jesus—who is radically different from them all, and whose justice aims to rescue and restore genuine humanness. The trouble is that in our flat-earth political philosophies we only know the spectrum that has tyranny at one end and anarchy at the other, with democracies as our dangerously fragile way of warding off both extremes. The news of God's sovereign rule inevitably strikes democrats and not just anarchists as a worryingly long step toward tyranny, as we apply to God and the Gospels the hermeneutic of suspicion that we rightly apply to anyone in power who assures us that they have our best interests at heart. But the story told by the Gospels systematically resists this deconstruction, for three reasons to do with the integration of the Gospel stories both internally and externally.

First, the narrative told by each Gospel—yes, in different ways, but at this point the canonical Gospels stand shoulder to shoulder against the so-called Gospel of Thomas and other similar documents—presents itself as an integrated whole in a way that scholarship has found it almost impossible to reflect. Attention has been divided, focusing *either* on Jesus's announcement of the kingdom and the powerful deeds—healings, feastings, and so on—through which it is instantiated *or* on his death and resurrection. The Gospels have thus been seen either as a social project with an unfortunate, accidental, and meaningless conclusion, or as passion narratives with extended introductions. Indeed, some methods of Gospel scholarship that have been developed over the years, so far from being neutral and objective, were designed to reinforce one or the other of these models. And even redaction criticism and composition criticism, which have tried to look at the Gospels as wholes, have found it difficult to see the big picture because the Western theological context has been so deeply resistant to the notion of God in public.

Thus the Gospels, in both popular and scholarly readings, have been seen either as grounding a social gospel whose naïve optimism has no place for the radical fact of the cross, still less the resurrection—the kind of naïveté that Reinhold Niebuhr regularly attacked—or as merely providing raw historical background for the developed, and salvific, Pauline gospel of the death of Jesus. If you go the latter route, the only role left for the stories of Jesus's healings and moral teachings is, as it was for the great German Rudolf Bultmann, as stories witnessing to the church's faith or, for his fundamentalist doppelgängers, stories that proved Jesus's divinity, rather than launching any kind of program (despite Luke 4, despite the Sermon on the Mount, despite the terrifying warnings about the sheep and the goats!).

Appeals for an integrated reading of the Gospels have met stiff opposition from both sides. Those who emphasize Jesus's social program lash out wildly at any attempt to highlight his death and resurrection, as though that would legitimate a fundamentalist program, either Catholic or Protestant, while those who emphasize his death and resurrection do their best to anathematize any attempt to continue Jesus's work with and for the poor, as though that might result in justification by works, either actually or at the existentialist metalevel of historical method. And these debates, which play out in supposedly neutral methods of Gospel study, simply reflect, however scholarly their proponents, the sterile antithesis between the twin tyrannies of secularism and fundamentalism. An integrated reading of the Gospels as they stand rejects both, offering not just a picture of what it would look like if the true God were running the world but the story—the history—of how that actually began to take place.

The lesson is this: (1) Yes, Jesus did indeed launch God's saving sovereignty on earth as in heaven, but this could not be accomplished without his death and resurrection. The problem to which God's kingdom project was and is the answer was deeper than could be addressed by a social program alone. Equally, (2) yes, Jesus did (as Paul says) die "for our sins," but his agenda of dealing with sin and its effects and consequences was never about rescuing individual souls *from* the world but about saving humans so that they could become part of his project of saving the world. "My kingdom is not *from* this world," he said to Pilate; had it been, he would have led an armed resistance movement like other worldly kingdom prophets. The kingdom he brought was emphatically *for* this world, which means that God has arrived on the public stage and is not about to leave it again, defeating the forces *both* of tyranny *and* of chaos—both of shrill modernism and fluffy post-modernism, if you like!—and establishing in their place the rule of restorative, healing justice, which needs translating into scholarly method if the study of the Gospels is to do proper historical, theological, and not least political justice to the subject matter.

The saying about "render unto Caesar," which is still regularly employed (as in the *Economist* feature!) to insist on the separation of God and public life, tells in fact heavily in the opposite direction, not least through its contextualization within Jesus's kingdom announcement. It is in the entire Gospel narrative, rather than any of its fragmented parts, that we watch the complete, many-sided kingdom work taking shape. And—my overall point—this narrative, read this way, resists deconstruction into power games precisely because of its insistence on the cross. The rulers of the world behave one way, declares Jesus, but you are to behave the other way, *because* the Son of Man came to give his life as a ransom for many. We discover that so-called atonement theology *within* the statement of so-called political theology, and to state either without the other—as both scholars and preachers have relentlessly done—is to resist the integration, the God-in-public narrative, which the Gospels, seen as integrated wholes, persist in presenting.

Second, the Gospels as wholes demand to be read in deep and radical integration with the Old Testament (a phrase I use advisedly because it reflects the perspective of the Gospels themselves, that they are bringing to its climax the great unfinished narrative of the Hebrew and Aramaic canon). Recognition of this point has been obscured by perfectly proper post-Holocaust anxiety about apparently anti-Jewish readings. But we do the Gospels no service by screening out the fact that each in its own way (as opposed, again, to the Gospel of Thomas and the rest) *affirms* the God-givenness and God-directedness of the Jewish narrative of creation, fall, Abraham, Moses, David, and so on, seen as the narrative of the creator God's rescue of creation from its otherwise inevitable fate. The Gospels claim that it was *this* project that was brought to successful completion in and through Jesus. The Gospels, like Paul's gospel, are to that extent folly to pagans, ancient and modern alike, and equally scandalous to Jews. We gain nothing exegetically, historically, theologically, or I suggest politically, by trying to make them less Jewishly foolish (or vice versa) to paganism and hence less scandalous, in their claim of fulfillment, to Judaism.

Third, the Gospels thus demonstrate close integration with the genuine early Christian hope. This, as I have argued elsewhere, is precisely *not* the hope for heaven in the sense of a blissful disembodied life after death in which creation is abandoned to its fate, but rather the hope, expressed in Ephesians 1, Romans 8, and Revelation 21, for the renewal and final coming together of heaven and earth, the final consummation precisely of God's project to be present, as Savior, in an ultimate public world. And the point of the Gospels is that with the public career of Jesus and his death and resurrection, this whole project was decisively inaugurated, never to be abandoned.

From the perspective of these three integrations, which I believe can be argued in great detail though not here, we can see how decisively mistaken are those readings, whether of the neognostic movement that is rampant today or the fundamentalism that is its conservative analogue. Indeed, if an outsider may venture a guess, I think the religious Right in the United States (we have really no parallel in Britain) may be construed as a clumsy attempt to recapture the coming together of God and the world that remains stubbornly in scripture but which the Enlightenment repudiated, and which fundamentalism continues to repudiate with its dualistic theology of rapture and Armageddon. It is as though the religious Right has known in its bones that God belongs in public but without understanding either why or how that might make sense; while the political Left in the United States, and sometimes on both sides of the Atlantic, has known in its bones that this God would make radical personal moral demands as part of his program of restorative justice, and has caricatured his public presence as a form of tyranny to evoke the cheap-and-gloomy Enlightenment critique as a way of holding that challenge at bay.

In particular, you might expect me to stress that the resurrection of Jesus is to be seen, through careful exegesis of the Gospels, not as proof of Jesus's uniqueness, let alone his divinity; certainly not as the proof that there is life after death, a heaven and a hell (as though Jesus rose again to give prospective validation to Dante or Michelangelo!), but as the launching *within the world of space, time, and matter* of that God-in-public reality of new creation called God's kingdom, which, within thirty years, would be announced under Caesar's nose openly and unhindered. And the reason why those who made that announcement were persecuted was because God acting in public is deeply threatening to the rulers of the world in a way that gnosticism in all its forms never is. The Enlightenment's rejection of the bodily resurrection has for too long been allowed to get away with its own rhetoric of historical criticism—as though nobody before Gibbon or Voltaire had realized that dead people always stay dead!—when in fact its non-resurrectional narrative clearly served its own claim to power, presented as an alternative eschatology in which world history came to its climax not on Easter but with Rousseau and Jefferson, with the storming of the Bastille and the American Declaration of Independence.

My central proposal, therefore, at an exegetical level, is that we offer a *political* hermeneutic of suspicion to those methods of

Gospel scholarship that have dominated much of the guild for the last few generations, to allow the political *meaning* of the Gospels—which is not to be played off against their theological meaning but rather integrated with it, if we are to be true to their intention—to emerge into the light of day. I remember being shocked and excited when a historian friend pointed out that Bultmann's agenda in the 1920s—to move away from Jesus as a hero figure and concentrate on the early community and its faith—was the direct correlate of Germany's rejection of the Kaiser and the establishment of *die Gemeinde,* "the community," in the form of the Weimar Republic. We could do with a similar analysis of the political setting of all Gospel criticism, my own included.

We also need to develop methods of study that will allow the four Gospels to emerge again as what they really are: the story of God's public kingdom project, which summons every individual, because it summons the whole world, to repentance and faith. This means would-be theological interpretations that ignore the political dimension, *as well as* would-be political interpretations that ignore the theological dimension, are simply ruled out as naïve and anachronistic. A genuine historical criticism—as opposed to ideological criticism (whether of Right or Left, secularism or fundamentalism) that uses the noble adjective *historical* as a smoke screen behind which to advance its reductive agenda—will insist on the both/and, not the either/or.

Wisdom for the Rulers—and the Church

Near the heart of the early chapters of Acts, a prayer of the church facing persecution makes decisive use of one of the most obviously political of the Psalms (Acts 4:24–30). Psalm 2 declares that, though the nations make a great noise and fuss and try to oppose God's kingdom, God will enthrone his appointed king in Zion and thus call the rulers of the earth to learn wisdom from him.

This point, which brings into focus a good deal of Old Testament political theology, is sharply reinforced in the early chapters of the Wisdom of Solomon. As we know, the royal house of David failed in an interesting variety of ways. But the promise and the warning still stand. Psalm 2 also appears at the start of the Gospel narratives, as Jesus is anointed by the Spirit at his baptism. Much exegesis has focused on the Christological meaning of "son of God" here; I propose that we focus equally, without marginalizing the Christology, on the political meaning. The Gospels constitute a call to the rulers of the world to learn wisdom in service to the messianic son of God, and as such they provide the impetus for a fresh biblical understanding of the role of the rulers of the world and of the tasks of the church in relation to them.

It is noteworthy that the early church, aware of prevailing tyrannies both Jewish and pagan, and insisting on exalting Jesus as lord over all, did not reject the God-given rule even of pagans. This is a horrible disappointment to post-Enlightenment liberals, who would have much preferred the early Christians to have embraced some kind of holy anarchy with no place for rulers at all. But it is part of a creational view of the world that God wants the world to be ordered, not chaotic, and that human power structures are the God-given means by which that end is to be accomplished— otherwise those with muscle and money will always win, and the poor and the widows will be trampled on afresh. This is the point at which Colossians 1 makes its decisive contribution over against dualisms which imagine that earthly rulers are a priori a bad thing (the same dualisms, as I have been suggesting, that have dominated both the method and the content of much biblical scholarship). This is the point at which the notion of the common good, advanced afresh by the Roman Catholic bishops in the 1990s and now reemphasized (though I think without full clarity) by Jim Wallis, has its contribution to make.

The New Testament does not encourage the idea of a complete disjunction between the political good to be pursued by the church and the political good to be pursued by the world outside the church, precisely for the reason that the church is to be seen as the body through which God addresses and reclaims the world. Here, I think, the essentially anabaptist vision of Wallis and others may need to be nuanced with a more firmly creational theology. (I know there are many varieties of anabaptism, and I hope it's clear that I am in considerable sympathy with many of their emphases, but there comes a point when anabaptism holds back from the dangerous task of working *with* the world, which I believe is just as Christian an obligation as working *against* the world.)

So, to put this first point positively, the New Testament reaffirms the God-given place even of secular rulers, even of deeply flawed, sinful, self-serving, time-serving, corrupt, and idolatrous rulers like Pontius Pilate, Felix, Festus, and Herod Agrippa. They get it wrong and they will be judged, but God wants them in place because order, even corrupt order, is better than chaos. Here we find, in the Gospels, in Acts, and especially in Paul, a tension that cannot be dissolved without great peril. We in the contemporary Western world have all but lost the ability, conceptually as well as practically, to affirm simultaneously that rulers are corrupt and must be confronted and that they are God-given and must be obeyed. That sounds to us as though we are to affirm simultaneously anarchy and tyranny. But that merely shows to what extent our concepts have led us again to muzzle the texts in which both stand together. How can that be?

The answer comes in such passages as John 19 and 1 Corinthians 2 and Colossians 2. The rulers of this age inevitably twist their God-given vocation (to bring order to the world) into the satanic possibility of tyranny. But the cross of Jesus, enthroned as the true Son of God as in Psalm 2, constitutes the paradoxical victory by which rulers' idolatry and corruption are confronted and overthrown. And the result, illustrated in Colossians 1:18–20, is that the rulers are *reconciled*—in some strange sense reinstated as the bringers of God's wise order to the world, whether or not they would see it like that. This is the point where Romans 13 comes in, not as the validation of every program that every ruler dreams up, certainly not as the validation of what democratically elected governments in one country decide to do against other countries, but as the strictly limited proposal, in line with Isaiah's recognition of Cyrus, that the creator God uses even those rulers who do not know him personally to bring fresh order and even rescue to the world. This, too, lies behind the narrative of Acts.

But this propels us forward into the perhaps unexpected and certainly challenging reflection that the present political situation is to be understood in terms of the paradoxical lordship of Jesus himself. From Matthew to John to Acts, from Colossians to Revelation, with a good deal else in between, Jesus is hailed as *already* the lord both of heaven and earth, and in particular as the one through whom the creator God will at the last restore and unite all things in heaven and on earth. And this gives sharp focus to the present task of earthly rulers. Before the achievement of Jesus, a biblical view of pagan rulers might have been that they were charged with keeping God's creation in order, preventing it from lapsing into chaos. Now, since Jesus's death and resurrection (though this was of course anticipated in the Psalms and by the prophets), their task is to be seen from the other end of the telescope. Instead of moving forward from creation, they are to look forward (however unwillingly or unwittingly) to the ultimate eschaton. In other words, God will one day right all wrongs through Jesus, and earthly rulers, whether or not they acknowledge Jesus and the coming kingdom, are entrusted with the task of *anticipating* that final judgment and final mercy. They are not merely to stop God's good creation from going utterly bad. They are to enact in advance, in a measure, the time when God will make all things new and once again declare that it is "very good."

All this might sound simply like cloud-cuckoo-land. It is bound to be seen as such by those for whom all human authorities are tyrants by another name. However, the Christian vision

of God working through earthly rulers makes the sense it makes
only if the church embraces the vocation to remind the rulers of
their task, to speak the truth to power, and to call authorities to
account. We see this going on throughout the book of Acts and
on into the witness and writings of the second-century apologists. And indeed, in the martyrs, because martyrdom (which is
what happens when the church bears witness to God's call to the
rulers and the rulers shoot the messenger) is an inalienable part of
political theology. You can have as high a theology of the God-
given calling of rulers as you like, as long as your theology of the
church's witness and martyrdom matches it stride for stride.

This witness comes into sharp focus in John 16:8–11. The
Spirit, declares Jesus, will convict the world of sin, righteousness, and judgment—of judgment, because the ruler of this world
is judged. How is the Spirit to do that? Clearly, within Johannine theology, through the witness of the church in and through
which the Spirit is at work. The church will do to the rulers of
the world what Jesus did to Pilate in John 18 and 19, when he
confronted the ruler with the news of the kingdom and truth,
deeply unwelcome and indeed incomprehensible though both
were. And part of the way in which the church will do this is by
getting on with and setting forward those works of justice and
mercy, of beauty and relationship, which the rulers know in their
bones ought to be flourishing but which they seem powerless to
bring about. But the church, even when faced with overtly pagan
and hostile rulers, must continue to believe that Jesus is the lord
before whom they will bow and whose final sovereign judgment
they are called to anticipate. Thus the church, in its biblical commitment to "doing God in public," is called to learn how to col-
laborate without compromise (hence the vital importance of the
theory of the common good) and to criticize without dualism.

In particular, it is vital that the church learn to criticize the
present workings of democracy. I don't just mean that we should
scrutinize voting methods, campaign tactics, or the use of big

money in the electoral process. I mean that we should take seriously the fact that our present glorification of democracy emerged precisely from that Enlightenment dualism, the banishing of God from the public square and the elevation of *vox populi* to fill the vacuum, which we have seen to be profoundly inadequate when faced with the publicness of the kingdom of God.

And we should take seriously the fact that early Jews and Christians were not terribly interested in how rulers came to be rulers—that is, the process by which they came to power—but they were extremely interested in what rulers did once they had obtained power. The greatest democracies of the ancient world, those of Greece and Rome, had well-developed procedures for assessing their rulers once their term of office was over, if not before, and if necessary for putting them on trial. Simply not being reelected (the main threat in today's democracies) was nowhere near good enough. When Kofi Annan retired as general secretary of the United Nations, one of the key points he made was that we urgently need to develop ways of holding governments, especially powerful governments, to account. That is a central part of the church's vocation, which we should never have lost and desperately need to recapture.

All this demands that the church be continually called to account, since we in turn easily get it wrong and become part of the problem instead of the solution. That is why the church must be *semper reformanda* as it reads the Bible, especially the Gospels. Fortunately, that's what they are there for, and that's what they are good at, despite generations of so-called critical methods that sometimes seem to have been designed to prevent them from being themselves. Part of the underlying aim of this is then to encourage readings of the Bible that, by highlighting the publicness of God and the gospel, set forward reforms that will enable the church to play its part in holding the powers to account and thus advancing God's restorative justice.

God in the Public World of Post-Postmodernity

Our culture is moving in all kinds of ways toward a post-post-modernism that has yet to be shaped but for which our public world longs as it lurches from boredom and trivia to dangerous and dehumanizing behavior. I have argued that the God of the Bible, and especially of the Gospels, can be understood only as God-in-public, and that methods of criticism designed to keep this rumor quiet need to be challenged by appropriate historical, theological, and political critique and replaced by methods that do justice to the reality of the texts and hence do justice—in the much fuller sense—in the public world that the Gospels demand to address. We face a challenging possibility in our generation: to move beyond the sterile alternatives of different types of post-Enlightenment tyranny on the one hand—the fundamentalisms and secularisms that have so often slugged it out on the spurious battleground of ideologically driven would-be exegesis—and postmodern chaos on the other. How might we do this?

There are many ways in which tomorrow's church could lead the way to a healthy and fruitful post-postmodernity. One of them is surely this: to take seriously the biblical witness to God in public, to think through its implications for the structure of political institutions as well as the decisions they need to make and, not least, to develop wise methods of exegesis that will allow the full flavor, wisdom, and challenge of the Bible to be heard again after the shrill certainties of fundamentalism, the equally shrill denials of secularism, and the incipient nihilism of the nevertheless necessary postmodern reaction.

Please note carefully: because this is explicitly an agenda of the common good, my proposal is not that biblical exegesis belongs to the church or to Christian scholars. Wise, holistic exegesis is a 'common' good project, and in turn it serves the larger common good projects that our world so badly needs. That is part of the creational and new-creational vision which, I have argued, is necessary if we are to read the Bible for all it's worth, and so to work with the grain of the God who made the public world in the first place and, despite rumors to the contrary, has never left it. Doing business with God in public is always complicated, but it is never dull.

10

How to Engage
Tomorrow's World

THE FAULT LINE between Christ and culture has been there ever since Jesus's public career, and particularly since Easter. Jesus fulfilled Israel's age-old hope, but not in the way anyone expected or wanted. The fault lines apparent during his public career were even more marked after his resurrection, and his first followers had to struggle hard, as we see in Paul, to figure out what following him would mean in relation to the traditional Jewish way of life.

But that struggle was small compared to the larger one. If Israel's Messiah is the lord of the world, then the line between Christ and culture, between the Messiah and the mores of the world, separates not simply two ways of life but the ambitious and powerful pagan world as a whole from the human embodiment of Israel's God, the man who claimed to be the rightful lord and master of that world. It isn't simply a matter of navigating between competing pressures. It's about being loyal to Jesus *as* the true lord of the world and believing, in consequence, that his way of life is what the world most truly needs. It is, in other words, an agenda for mission and service. But since Jesus's way of life is the path of self-giving love, that mission and service can never be about imposing a would-be Christian policy or ethic on an unwilling or unready public, but rather allowing Jesus's way of bringing his kingdom to work through us and in us. The church at its best has always sought to transform society from within.

Early Christian answers shifted this way and that, partly in response to the position taken by the local culture. When Paul was in Greece, a leading magistrate ruled that Christianity was an internal division within Judaism (Acts 18:12–16). Since Judaism was legally permitted, there was no problem with Christianity either. The Greek church was thus able to grow without state interference. However, in Asia Minor (modern Turkey) there was a problem. Most Christians were former pagans, and following Jesus meant no longer worshipping the local deities. And the local deities now included the imperial cult. The Jewish people had official exemption from pagan worship, including imperial worship. But when the Christians also opted out, the trouble started. Pressure was put on the Jewish community, as well as Jewish Christians, to bring the ex-pagans into line by having them circumcised. It's complicated, isn't it?—but no more complicated than the situation of many Christian citizens and institutions in our own world. Where do you draw the line? What counts as compromise? Take heart: these questions are not new. They are as old as the gospel. The fact that you face them today doesn't mean you've taken a wrong turn somewhere. This is Christian normality. And Jesus remains sovereign over it.

In fact, by struggling with those questions, the early church was living out its own version of what the Jewish people had been doing ever since the exile. They moved to and fro along a spectrum. Sometimes their pagan rulers were helpful, such as when Cyrus gave the command to rebuild the temple. Sometimes they weren't, as when Antiochus Epiphanes tried to stamp out the Jewish food laws or when Hadrian tried to ban circumcision. The book of Daniel shows the full range, from violent persecution to promotion within the civil service. The questions then press thick and fast, as they did for both Jews and Christians in the pagan diaspora: What counts as loyalty and disloyalty? What counts as dangerous compromise, and what as wise flexibility? When do you resist, when do you run away, when do you stay and try to improve things from within, and when do you stay and face martyrdom? How do you know, and who says?

The New Testament as a whole suggests that it is part of normal Christian life to face these questions. A quick flip through 1 Peter or Revelation indicates that the Christian's normal state is to be out of sorts with rulers and authorities, which is why it's all the more important, as Peter insists, that Christians be blameless in all things except their faith and what it requires. If they're going to attack us, at least let them attack us—like Darius's men with Daniel—for the right reasons. And when we get into the second, third, and fourth centuries, we find a wide variety of stances, as the church made its way deep into pagan society, sometimes incurring massive opposition, persecution, torture, and martyrdom.

So what are the starting points for wise Christian reflection on where we are and how we can find the courage to face the confusions and challenges of tomorrow's world? In addition to Jesus's achievement, whose consequences I tried to spell out in my book *Simply Jesus,* I want to touch on two passages: Paul's letter to the Colossians and the Gospel of John.

Christ, Culture, and Paul's Letter to the Colossians

Paul's letter to Colossae carries an explosive charge in inverse proportion to its short length. Paul wrote it from prison. When he makes his sweeping and all-embracing statements about Jesus, he knows he is flying in the face of the apparent reality. And yet he doesn't hold back:

> He is the image of God, the invisible one,
> The firstborn of all creation.
> For in him all things were created,
> In the heavens and here on the earth.
> Things we can see and things we cannot—
> Thrones and lordships and rulers and powers—
> All things were created both through him and for him.
>   (Colossians 1:15–16)

And so on. We need to let Paul remind us, precisely when major cultural change is upon us, that our confidence is not in the solidity of Western culture or the basic goodness of modern democracy. Our confidence is in Jesus and him alone. We need this message with every fiber of our beings. And we need to generate and sustain communities and educational institutions where this extraordinary, breathtaking, ridiculous truth is woven into the very fabric of all we do.

Some might see this great claim as shutting down all academic inquiry: If Christ is the answer, what's the point of asking the question? I see it exactly the other way around: Jesus is lord of the world, so all truth is his truth; let's go and explore it with reverence and delight. Whether you look through the telescope or the microscope, whether you study texts or traditions, whether it's oceanography or paleography, you are thinking Jesus's thoughts after him.

In particular, Paul declares—despite languishing in prison!—that all rulers and authorities were created through and for Jesus the Messiah. We are called today to think afresh what that might mean in terms of modern Western democracy. What has Colossians 1 to do with the present wrangling in the European Union and Britain's edgy position in relation to it? What, to take another random example, has Colossians to do with primary polls and party structures, expensive advertising and showpiece debates? For Paul it wasn't a matter of looking at a particular political structure and assessing how well or badly it might serve God's purposes. This was an apriority. God made the world to be looked after by human beings; that's part of being made in God's image. Even though humans rebelled, God still intends to bring order and harmony to his world through human agents—though, because they are still rebels, that order and harmony often becomes sterile, dehumanizing tyranny.

God takes that risk. But within this world of rebels and risks, God has established in the Messiah, Jesus, a new people who are his body, the place where his glory already shines in the world. The church is not simply a religious body looking for a safe place to do its own thing within a wider political or social world. The church is neither more nor less than people who bear witness, by their very existence and in particular their holiness and their unity (Colossians 3), that Jesus is the world's true lord, ridiculous or even scandalous though this may seem.

The danger, as we well know, is that though we in the West have retained our distinctive Christian witness in some areas, we have undoubtedly compromised it in others. It is part of the basic Enlightenment settlement, to which the United States gave explicit allegiance in the Declaration and Constitution and to which Britain gives backhanded allegiance in a thousand unwritten ways, that the church should step back from public life and do its own thing in private.

This reflects the latent Epicurean philosophy, which has been a feature of mainstream Western thought since at least the fifteenth century with a crescendo in the eighteenth and nineteenth: God, or the gods, are a long way away, and the world will get on under its own steam. Within that worldview, the church can purchase its independence by colluding with the implied pagan philosophy. All right, we say, if God and the world are split so far apart, we'll just do the God bit, giving people a private spirituality in the present and a blissful hope for the future, but not engaging in radical questioning of the systems that result. I suspect that one of the reasons why the creation/evolution debate generates so much heat in America—far more than anywhere else—is that people can hear all the overtones, social, cultural, and political, that it throws off. The idea of God having anything to do with the ongoing process of the world flies in the face of all that Western culture has stood for—including Western Christian culture.

But Paul would have nothing to do with that. He was in prison precisely because he had announced, and taught people to live by, a message in which Jesus claimed the ultimate rights over every aspect of life. C. S. Lewis famously summed this up by saying that there was no neutral ground in the universe. Every square inch, he wrote, every split second, is claimed by God and counterclaimed by Satan. Lewis was echoing the view of many Christian thinkers, going back to Abraham Kuyper and ultimately to Paul himself; but in doing so, he, and they, stand firmly against the great division that has come upon us in the West.

Taking this stand is dangerous. Attacks do not usually come head-on. Satan regularly disguises himself as an angel of light. When a church or a Christian organization hears the reminder that it is dealing with spiritual matters and must back off and let the world run everything else, we may not glimpse the horns, the hooves, and the pointy tail, but the challenge to the lordship of Jesus the Messiah is nonetheless real. I ask myself, and I ask you, whether the problems of this new generation, with governments pushing unwelcome agendas on us, may not be a new manifestation of something with which we have colluded for many years. Have we fostered a culture in which the lordship and teachings of Jesus, for instance about poverty or human dignity or war, have been honored, studied, taught, and practiced? Or have we been content—as so many Christians on both sides of the Atlantic have been content—to drift with this or that prevailing political wind, to trim our sails so that only one or two real distinctives are left, related perhaps to sexual and family life, only then to complain when the principalities and powers, having quietly gained our cooperation in other spheres, such as rampant individualism and the neoliberal vision of the good life that goes with it, now come to attack those last remaining strongholds?

Colossians, then, insists upon the total and supreme lordship of Jesus the Messiah, not by saying that other authorities do not exist but by saying that those which do are subject to his lordship. But what does that subjection look like? How does it come about? It comes about through those who belong to the Messiah, who have died and been raised with him, lifting up their heads and their hearts and learning how to live that hugely attractive life, clothed with the garments of Jesus himself, a transformation that has always made even the most hardened pagan observers gasp and wonder how it's done.

The pagan world is described in Colossians 3, in two lists of pagan behaviors: sexual immorality in all its forms on the one hand, and malicious and angry speech on the other. Oh, the irony of it. I know many churches that wouldn't tolerate the slightest sexual immorality (at least not outwardly) but which are hotbeds of gossip, backbiting, and malice. And I know other churches where everyone is nice to everyone else, but sexual license reigns unchecked and everyone is "supportive" about it, because otherwise one would be "saying nasty things." By contrast, I have often thought that the alternative model Paul proposes commends itself as the kind of society everyone would really like to belong to:

> You must be tender-hearted, kind, humble, meek, and ready to put up with anything. You must bear with one another and, if anyone has a complaint against someone else, you must forgive each other. . . . On top of all this you must put on love, which ties everything together and makes it complete. . . . And whatever you do, in word or action, do everything in the name of the master, Jesus, giving thanks through him to God the father. (Colossians 3:12–17)

These are the positive virtues of the Christian community. They don't happen by accident. You have to think about them, individually and corporately. You have to work at them together, to repent of failures in these areas, and to forgive one another. And when we do that, bad behavior of body or tongue will be shown up as what it is: socially destructive and dishonoring to God the creator.

Communities like this are the way God changes the world—not by retreating from the world but by going boldly, as Paul did, into the places of power and authority in the world, praying for a door to be opened for him to speak about the mystery of the world's true king (4:3). That, as he says, is why he was wearing chains when he wrote this letter. But the gospel was not chained up. It was out there, transforming lives and communities under the rule of the one true sovereign. Here is the heart of our confidence, whatever the world around may throw at us: the crucified and risen Jesus is *already* lord of the world. This is a call to renewed prayer, to renewed holiness, to renewed and cheerful confidence in the power of Jesus to make a fresh way forward.

The Spirit and the World

If one cannot do justice to Colossians in a few pages, how much less can one deal briefly with John's Gospel. I want simply to draw out one strand that is not always highlighted but nevertheless is a word for our time. We often read John through the darkened spectacles of nineteenth-century pietism and thereby miss the bright, sharp Johannine light on the church's social and political witness. One of the major problems of our world is that the church has colluded with two major shifts that occurred in the eighteenth and nineteenth centuries. We haven't tried to put it right because we haven't even noticed. A word about that double shift, and then a look at John.

From the eighteenth century on, the state has been taking over things the church used to do. Hospitals, schools, and the like were Christian innovations. Outside Christianity, you got medicine and education if you could pay, and you suffered in ignorance if you couldn't. The transformative genius of the early Christian movement was to embody the outgoing, practical love of God in Jesus the Messiah for all people. Nobody had ever thought of doing it that way before, but it caught on and was a primary reason for the church's rapid spread. To this day, a Jewish friend of mine reports that when her children babysit for Christian couples who are out volunteering in the youth club or the prison or the hospice, the children simply say that the couple are out "being Christian." Thank God that is how the church is still sometimes seen. And thank God for the basically Christian impulse that has led the modern state to see that medicine and education are for all, not only for the wealthy. But when the state then claims the right to dictate how that is to be done and tells the church to back off and stick to private spirituality and not interfere with debates about genetic engineering or assisted suicide, or how we teach history or economics or art, we must hold our ground. That stuff is part of our core vocation; we are delighted that whole societies now want to share it, but we are not going to give it up or be told how to do it. How we say that is obviously a matter of wisdom, of gracious speech seasoned with salt. That we say it should be nonnegotiable.

This leads to the second point. From pre-Christian Judaism to the present, God's people have claimed the right and responsibility to speak truth to power, sometimes with words, often with bodies. Martyrdom has frequently been the most powerful statement of all. The post-Enlightenment world has developed two other ways of speaking truth to power, but neither has done the job as well.

On the one hand, we have opposition parties, which easily generate a two-party culture-war polarization, which both Britain and the United States suffer from. Every issue is seen in black-and-white terms of us and them. If that doesn't happen, you get a riot of small parties generating unstable coalitions and shady deals irrelevant to the real issues. That's the problem with oppositional democracy.

On the other hand, we have the electronic and print media, the increasingly complex world of journalism that takes on itself the responsibility of holding government, and indeed the opposition, to account. If you doubt this, have a look at the quotations carved in stone in the lobby of the Tribune Building in Chicago. Take this, from Robert R. McCormick, owner and publisher of the *Chicago Tribune* eighty years ago: "The Newspaper is an institution developed by modern civilization to present the news of the day, to foster commerce and industry, to inform *and lead* public opinion, *and to furnish that check upon government which no constitution has ever been able to provide*" (italics added). That is breathtaking, all the more because that is *our* job as the followers of Jesus. (I can't resist adding that McCormick famously carried on campaigns against—and I quote shamelessly from Wikipedia's entry on McCormick—"gangsters and racketeers, prohibition and prohibitionists, local, state, and national politicians, Wall Street, the East and Easterners, Democrats, the New Deal and the Fair Deal, liberal Republicans, the League of Nations, the World Court, the United Nations, British imperialism, socialism, and communism." So it wasn't only governments he had in his sights.)

These two methods of speaking truth to power—official opposition parties and the media—regularly fail. As we all know, opposition parties often collude with governmental folly and wickedness, and newspapers can easily egg them on in precisely those areas where critique is most needed. The church's vocation of speaking truth to power has thus been taken over by two systems that aren't up to the job. We urgently need the voice of Christian wisdom to approve that which is excellent and to call to account that which isn't. Of course, when we try to do that, the media regularly tries to rule the church out of order, not just because it doesn't like what we might say but because we are treading on turf they took from us, and they don't want us to have it back. So, once again, we have colluded with this diminishing of our historic role and God-given vocation; or, worse, we have been herded like sheep into the lobby of this or that party, swept along on agendas we assume too readily to be God's agendas and unable to differentiate between the whim of the party and the conscience of the Christian.

So what has John's Gospel got to say on this? It gives us Jesus's description of what will happen when the Holy Spirit is given to his followers:

> When [the Spirit] comes, he will prove the world to be
> in the wrong on three counts: sin, justice, and judgment.
> In relation to sin—because they don't believe in me. In
> relation to justice—because I'm going to the father, and
> you won't see me any more. In relation to judgment—
> because the ruler of this world is judged. (John 16:8–11)

These verses feel dense and obscure. But in the larger context they are all too clear. Jesus has already spoken of his overthrow of "the ruler of this world," as a result of which he will now "draw all people to myself" (12:31–32). He has warned that the world will hate his followers as it hated him, and he is about to tell them that they should cheer up because he has overcome the world (15:18–25; 16:33). And at the climax of the book in chapters 18 and 19, Jesus confronts Pontius Pilate—God's kingdom confronting Caesar's kingdom—and explains to him that God's kingdom is based on truth, a truth that redefines power. That redefinition is then put into practice, as Pilate, weak and vacillating, sends him to the death by which he, sovereign and powerful, completes the work of the world's redemption (19:30). The hidden truth, the hidden power, the energy that drives God's kingdom, is divine self-giving love: having loved his own in the world, he loved them to the end (13:1).

So what does Jesus mean about the Spirit proving the world wrong? And how is this to happen? Christians of recent generations have spoken of the Spirit in terms either of miraculous inspiration or of delightful personal experiences. But when the Spirit comes to do what Jesus is describing here, the church will not sit back watching it happen. No, the way the Spirit will prove the world to be in the wrong is *through us and our witness.* This is part of our core vocation. Not to do it is to quench the Spirit.

Take the three elements one by one. The church, praying for the Spirit, must first embrace the vocation to show up the world in relation to sin, because in failing to believe in Jesus the world is missing what genuine human life is like. This outward-looking vocation is another reason why it is vital that the church maintain its personal and corporate discipline, because tragically the world has had little difficulty finding accusations to throw back at the church. Of course, the way the church is supposed to show up the world is not by a sneering, holier-than-thou attitude but by providing such a wonderful model of God's genuine humanity that the world is seen as sordid and shabby in contrast, a place of lies and death instead of truth and life.

Second, the church is to prove the world wrong about justice. Jesus is to be vindicated by the father in his ascension, and this is the ultimate moment of justice, of putting the world right. The world thinks it knows what justice is, but again and again the world gets it wrong, favoring the rich and powerful, turning a blind eye to wickedness in high places, forgetting the cry of the poor and needy who the Bible insists are the special objects of God's just and right care. So the church, in the power of the Spirit, has to speak up for God's justice, in the light of Jesus's ascension to the throne of the world, and to draw the world's attention to where it's getting this wrong.

This has immediate and urgent application in holding our governments to account concerning justice for the world's poorest, who have been kept poor by the unpayable compound interest owed to Western banks on loans made decades ago to corrupt dictators. The injustice has itself been compounded by our governments' breathtaking bailing out of superrich companies, including banks, when they defaulted: the very rich did for the very rich what they still refuse to do for the very poor. In the early church, bishops got a reputation for tirelessly championing the needs and rights of the poor. They were a nuisance to the rich and powerful, but they would not shut up. They were doing precisely what Jesus says would happen when the Spirit came and held the world's injustice up to the light of his justice. As well as investigating obvious injustices, the church has the responsibility to test those causes that claim the words *justice* or *rights* but are in fact merely special-interest groups. As Pope Benedict XVI said in his address to the United Nations in April 2008, the language of rights is borrowed from the great Christian tradition, but if you cut off those Christian roots, you get all kinds of abuses, each claiming the postmodern high ground of victimhood but only succeeding in debasing the coinage of rights itself. Part of the task of holding the world to account is thinking and speaking clearly, humbly, and wisely in these areas.

Third, the church in the power of the Spirit is to prove the world wrong about judgment. On the cross and in the resurrection, Jesus passed judgment on the dark power he calls "the ruler of this world." The dark lord operates through violence and death and the threat of both. Jesus takes their full force on himself and shows in the resurrection that he has overcome them. He has launched God's new creation, and the powers of death know that they are beaten. This puts in question all use of violence, all attempts to use death to control the world. There are many occasions when countries and governments need police; given the wickedness of the world, police sometimes need to use force. But this exceptional dispensation all too quickly becomes a lust for the power that bases itself on violence, and it is precisely that which the Spirit, through the church, must call to account.

I don't think we in the Western churches have sufficiently re-flected on these three sharp-edged vocations that Jesus bequeaths to his people by his spirit. We have colluded so much with the Enlightenment's shrinking of the church's role that we are likely to react with surprise or alarm to such suggestions. Yet there they are in scripture. My point is this: How do we shape a generation through which the Spirit will convict the world of sin (in the face of Western arrogance and assumed moral superiority), of justice (in a world where biblical meaning, justice for the poor, has been obliterated by justice in the shape of state-sanctioned violence), and of judgment (in a culture that acts as if it were the arbiter of truth)? That is the challenge.

There will be enormous resistance to this, within the church as well as outside. We would be claiming back ground that we've not only lost but in most cases have forgotten we ever possessed. What's more, we would be courting martyrdom of one sort or another. Jesus, going to his death, told us that servants are not above their master. But that simply puts us back in the center of the map of Christ and culture that I sketched earlier. Martyrdom won't always happen. Sometimes, perhaps often, in the mercy of God and the power of the Spirit, the church's witness will lead to a major change of heart. That has happened before; because of the resurrection, I believe it can and will happen again. But it isn't new. It's what Jesus not only promised but also mandated and modeled.

IN JOHN'S HAUNTING AND suggestive closing chapter, we eaves-drop on one of the most embarrassing encounters in the whole Bible, as Jesus three times asks Peter if he loves him. All Chris-tian leaders at some time or another cling to that conversation for dear life, because we know in shame that we have let our Lord down. All of us have heard the astonishing word of Jesus, "feed my lambs," "look after my sheep," "feed my sheep": words of commission that are also words of forgiveness, and vice versa.

The fresh personal meeting with the Lord himself, the frank and humble confession, and the fresh commission—this is what we as Christian leaders live on day by day and week by week.

But my theme here takes us to the next and final scene of that same chapter. When we face new challenges, we are inclined, as Peter was, to look around and see who else is out there and what they're doing. "Lord," we say, "what about this man?" I believe we should, individually and collectively, take Jesus's answer as the signal for fresh obedience to the challenges we face. Other organizations may do it differently. Other individuals may do it differently. Our task is not to think about them, but to think about him and what it will mean to follow him into a future that is unknown to us but held firmly under his victorious sovereignty. Jesus turns to Peter. "If it's my intention," he replies, "that he should remain here until I come, what's that got to do with you? You must follow me!" We never know what situations we will meet around the corner. But that isn't the source of Christian hope or Christian confidence. Our hope and confidence come from knowing who it is we are following. "What's that got to do with you?" he asks. Other people's problems and challenges are other people's problems and challenges. "You must follow me."

11

Apocalypse and the Beauty of God

I LAY BEFORE you three puzzles, and I'm going to suggest that the way to solve all three is to put them together and allow them to solve one another. The first is that the notion of apocalyptic, of a great cataclysm through which God's ultimate purpose will come to fruition, has become in our day the weapon—the metaphor is apt—of a particular and highly influential school of thought, not least in America. The puzzle of apocalyptic, for any serious Christian, any thoughtful reader of the New Testament, is whether, and if so how, apocalypse can be rescued from the "left behind" school of thought, whose adherents anticipate the rapture in which they will be snatched up to heaven, leaving this world behind once and for all. Those who take this view have no reason to worry about the condition of the present world or issues like global warming or acid rain; indeed, they sometimes take pride in pollution, since the world is not their home, they're just a-passing through, and if they can hasten its demise so much the better.

This careless attitude to creation goes with an eagerness for

war, especially certain types of wars with certain types of enemies, particularly the war that will lead to the great Armageddon.
This brand of theology has become highly influential, again especially in America. Can the whole biblical notion of apocalyptic
be rescued from those who read it this way? Or must we declare,
with the liberal theologians of yesterday and the liberal politicians
of today, that the whole thing is outdated nonsense, and if that's
what the Bible says, the less we read it the better?

My second puzzle is very different, and it has to do with the
place of the arts within a Christian worldview. I was brought up
in a world where the arts constituted, as it were, the pretty border
around the edge of reality rather than a window on reality. It was
nice to have good music, great art, fine architecture, even perhaps brilliant dancing, in society and the church, but within the
modernism of my youth it didn't seem to integrate with real life
in the world or with real Christian faith. The arts were recreation
and relaxation for those who liked that kind of thing, but (except
for dangerously subversive characters like playwrights) we didn't
expect them to impinge on how we organized the world, how
we ran the country, how we did our work, or indeed how we
understood and expressed our faith. I grew up singing Handel's
Messiah and Bach's *Saint Matthew Passion,* but I think I and my
contemporaries regarded the music as more or less a way of sugar-
ing the pill, of making the Bible listenable, and not as something
to be integrated more tightly within a Christian worldview.

I suspect that many in our world and our churches struggle
with this question, not least many whose talents lie in the arts
but who find that neither the world at large nor the church know
what to do with them, what they are (so to speak) there for. In
my experience, the Christian painter or poet, sculptor or dancer,
is regularly regarded as something of a curiosity, to be tolerated,
humored even, maybe even allowed to put on a show once in
a while. But the idea that they are, or could be, anything more
than that—that they have a vocation to reimagine and reexpress

the beauty of God, to lift our sights and change our vision of reality—is often not even considered.

At the same time—a further twist within my second puzzle— the world of postmodernity has made it harder still to see what the arts are there for or how they could be properly valued within the Christian community. Without going into detail, the eclecticism of postmodernity, starting with architecture but spreading rapidly elsewhere, and the sense that modernist pretensions have to be scorned and mocked have led to polarization within the arts with equal unhappiness at both ends. On the one hand, we have sentimentalism, making it difficult to say something positive or cheerful without appearing to collapse into kitsch. On the other hand, the late-modern brutalist movement in architecture has spread into a kind of in-your-face antiaesthetic, where the uglier, the more violent, the more shocking something is, the more it appears to be real art, thus denying the triviality, the mere prettiness, of older visions of beauty and seeking to draw attention to and even wallow in the horror and apparent meaninglessness of life. How can you be an artist—how can you be a *Christian* artist—when the culture is polarized in that way?

The third puzzle is an exegetical one. You will be familiar with the majestic scene in the sixth chapter of the prophet Isaiah, a passage often read at confirmations and particularly ordinations. In the temple, the prophet has a vision of YHWH surrounded by angels, with the house filled with smoke. The angels are singing a song that is echoed in Jewish and Christian liturgies to this day: "Holy, Holy, Holy is YHWH the God of hosts; the whole earth is full of his glory." *The whole earth is full of his glory.* That's a remarkable thing to say, and the prophet's instant reaction is to say, "No, it's not! There's me for a start: I am undone; I'm a man of unclean lips, living in amongst a people of unclean lips!" And the scene continues with the terrifying cleansing of Isaiah and his commissioning to speak powerful words of judgment and, ultimately, of mercy.

But the song of the angels in chapter 6 stands in tension with what we find in the same book just a few chapters later. In the vision of the future in chapter 11, the vision of God's anointed ruling in wisdom and justice, we find that great promise of creation restored: "The wolf shall live with the lamb, the leopard shall lie down with the kid, the calf and the lion and the fatling together, and a little child shall lead them. . . . They will not hurt or destroy on all my holy mountain; for the earth will be full of the knowledge of YHWH as the waters cover the sea." *The earth shall be full of the knowledge of YHWH,* a prophecy repeated in Habakkuk 2:14, *the earth will be filled with the knowledge of the glory of YHWH.* (See too Psalms 33:5; 72:19.) And we want to ask Isaiah: Well, what are you saying? Is the earth *already* full of YHWH's glory? Or is this something we have to wait for, until the future when everything has been set to rights?

Perhaps you already see how these three puzzles might begin to tie up—the question of apocalypse, the question of the role of the arts in general and particularly within the church, and this sharp-edged exegetical problem. Let me give you my proposal in miniature, and then spell it out in terms of a couple of majestic biblical passages: Isaiah 65:17–25 and Revelation 21:9–27. My proposal is this, beginning at the middle with the question of art. True art, I suggest, approximates more and more the vision of the way things are and the way things will be. We humans know in our bones that we are children of the present creation, which is simultaneously both glorious and shameful, and that we are designed for a fuller creation, a new order, a world flooded with the creator's glory, full of justice and joy and, yes, beauty. The point of new creation is that it is the redemption and transformation of this present creation, with its shame and horror overcome; that is the way, if I can put it like this, to the reconciliation of Isaiah's dilemma.

And the true point of biblical apocalypse, as opposed to the distorted and dualistic versions that have been so powerful and prevalent in our day, is that biblical apocalypse is all about God's future breaking into the present, seen in glimpses, known above all in Jesus, and best expressed not in abstract theology or even in preaching but, yes, in genuine and visionary art. Apocalypse, both in form and in biblical content, is not about denying the present creation but about overcoming its sorrows and realizing its promise. Apocalypse is the key to understanding and reexpressing the beauty of God.

Let me spell this out a little more fully. A fully biblical worldview requires that we hold tightly to three things in particular. First, the goodness and God-givenness of the present creation: the whole earth is full of YHWH's glory, and any attempt to suggest that the created order is bad or shabby is a denial of that glorious truth. But, second, as Isaiah protests, the world is also full of radical evil, of human wickedness and its fruits, and to deny that is to live in a sentimental cloud-cuckoo-land. Sometimes, as in gnosticism but not in scripture, this second truth is allowed to trump the first, so that the evil in the world blots out the recognition of goodness, of the presence of the creator's glory. But, thirdly and vitally, biblical books from Isaiah to Revelation, and not least the great New Testament theologians Paul and John, speak of new heavens and a new earth, the renewal and restoration of creation and not its abandonment. And when they do so, they speak in particular of the new Jerusalem—not, as in some would-be Christian imagination, a purely heavenly city that has left earth behind, but a city that comes down from heaven to earth, in the final fulfillment of Jesus's prayer. And part of the point of Jerusalem always was that it was the place where God's glory would fill the temple, as it filled the tabernacle in the wilderness, not in order that the rest of the world becomes irrelevant but rather in anticipation of the eventual promise of God's glory filling the whole earth (Exodus 40:34ff.; 1 Kings 6:11; 2 Chronicles 5:13ff., 7:1ff.).

This promise about the temple is repeated even when Jerusalem is under threat of imminent judgment (Ezekiel 10:4) and as part of its promise of restoration (Ezekiel 48:35; 2 Maccabees 2:8). The great promises in Isaiah return to this point: YHWH will do a new thing, remaking creation so that the desert blossoms like a rose, and then his glory will be revealed and all flesh will see it together (Isaiah 35:2, 40:5, 60:1). And it is out of this matrix of thought that apocalypse arises, not in itself the dualistic worldview sometimes imagined (though it can shade off into that where the writers' grip on the goodness of creation and the promise of new creation is weakened), but rather as a glimpse of the already existing reality of new creation from within the old, so that those living within the old catch sight of the new, inviting them not to escapism but to hope. Apocalypse thus resonates with the moment in the desert when Jacob dreamed of a ladder between heaven and earth and declared upon waking that YHWH was in the place and he hadn't known it.

The point of art, I believe, is not least to be able to say something like that, to draw attention—not to a shallow or trivial pietistic point, as though to lead the mind away from the world and its problems and into a merely cozy contemplation of God's presence, but rather to the multilayered and many-dimensioned aspects of the present world, to the pains and the terror, yes, but also to the creative tension between the present filling of the world with YHWH's glory and the promised future filling, as the waters cover the sea. When art tries to speak of the new world, the final world, in terms only of the present world, it collapses into sentimentality; when it speaks of the present world only in terms of its shame and horror, it collapses into brutalism. The vocation of the artist is to speak of the present as beautiful in itself but pointing beyond itself, to enable us to see both the glory that fills the earth and the glory that will flood it to overflowing, and to speak, within that, of the shame without ignoring the promise and of the promise without forgetting the shame.

The artist is thus to be like the Israelite spies in the desert, bringing back fruit from the promised land to be tasted in advance.

That story, indeed, is one of the moments when YHWH surprisingly promises that not only the promised land but the whole world will be filled with his glory (Numbers 14:21; cf. 14:10). But just as not all the spies brought back an encouraging report, many artists recoil from the challenging vision of the future and prefer to give the apparently more relevant message of despair. Here is the challenge, I believe, for the Christian artist, in whatever sphere: to tell the story of the new world so that people can taste it and want it, even while acknowledging the reality of the desert in which we presently live.

And so, with our puzzles beginning to discover some resolution by being brought together, we return to the two spectacular passages. Isaiah's vision of new heavens and new earth, drawing on the earlier vision of chapter 11, highlights the joy of the holy city as a place without crying or distress, a place of covenant blessing and renewal (65:21ff., echoing Deuteronomy 28:30ff.), a place above all of new harmony within the created order, with the wolf and the lamb lying down together and the lion becoming a vegetarian and eating straw. Part of the difficulty faced by those who have dreamed dreams of new countries, of new lands where all would be well, is that to arrive at this utopia they always seem to have to do some fairly unutopian things. As a global society, we are caught in exactly that bind, as we proudly announce that we believe in peace and freedom and, to prove the point, drop yet more bombs and keep yet more people enslaved in hopeless debt. The way we currently do global empire is in need of radical, indeed apocalyptic, critique. And perhaps it is not least artists who, if given the encouragement and support they need, can help us mount that critique, break through the postmodern barrier that stops us from glimpsing new truth, and point to a wiser and more fruitful way forward.

That great vision at the end of the Book of Revelation is a vision of ultimate beauty. The word *beauty* doesn't occur much in the Bible, but the celebration of creation all the way from Genesis, through the Psalms and prophets, on into the Gospels and here in Revelation, should alert us to the fact that, though the ancient Jewish people did not theorize about beauty like the Greeks did (that's another story, and a fascinating one, though not for today), they knew a great deal about it and poured their rich aesthetic sensibility not only into poetry but also into one building in particular: the temple in Jerusalem, whose legendary beauty inspired poets, musicians, and dancers alike. This is the temple where YHWH's glory is glimpsed, not as a retreat *from* the world but as a foretaste of what is promised *for* the whole world. In the great vision of John, the temple has disappeared because the whole city has become a temple; the point of the city is not that it is a place of retreat from a wicked world but that its new life is poured out into the whole world, to refresh and heal it.

Sadly, verses 15–21 of Revelation 21 are often omitted in public reading, presumably because those who compose lectionaries suppose these verses to be boring and repetitive. But in passages like this we see, with the eye of the apocalyptic visionary, the astonishingly powerful beauty of God's new creation, beauty that should serve as an inspiration to artists and, through their work, to all of us as we seek to give birth to the life of the new creation within the old. The golden city, perfectly proportioned, equal in length and breadth and even, remarkably, height, has, says John, the glory of God and a radiance like a very rare jewel, like jasper, clear as crystal. The wall is built of jasper, while the city itself is pure gold, clear as glass. The foundations are adorned with jewels: jasper, sapphire, agate, emerald, onyx, cornelian, chrysolite, beryl, topaz, chrysoprase, jacinth, and amethyst. The twelve gates are twelve pearls, while the streets of the city are pure gold, transparent as glass. I confess that my knowledge of jewelry is so poor that I can't at once envisage those shining foundations, but I know that whoever wrote this passage delighted in them and wanted readers to do the same, relishing them one by one and in their glittering combination.

I know too that the idea of city streets paved with gold had nothing to do with fabulous wealth—pity the poor human race, when dazzling beauty is reduced to purely monetary function!— but rather with the most ravishing and wonderful beauty imaginable. This is the apocalyptic vision of the beauty of God. And it is given to us not so that in desiring to belong to that city we forget the present world and our obligations within it, but so that we will work to bring glimpses of that glory into the present world, in the peacemaking that anticipates the Isaianic vision of the wolf, the lamb, and the vegetarian lion; in the doing of justice that anticipates the final rule of the true Messiah; in the work of healing that springs from the water of life flowing from the city into the world around; and not least in the glorious art that gives birth to genuine beauty within a world full of ugliness, which bridges the gap between Isaiah's present and future visions, a world full of glory and a world to be filled yet more completely.

As an aid to this reflection, and to the vocations that follow from it, let me close with a truly remarkable example of the sort of thing I mean. In Revelation 22, the river of life flows from the city to irrigate the surrounding countryside, and on its banks grows the tree of life: not a single tree, as in Genesis, but many trees, now freely available, bearing fruit each month and with leaves for healing. This image of the tree of life and the radical and beautiful healing it promises has generated an extraordinary work of art, commissioned jointly by the British Museum and Christian Aid, and created by artists in Mozambique after the end of that country's long and bitter civil war.

The work is a sculpture of the tree of life. It stands nine or ten feet tall, with branches spreading nine or ten feet in all directions. In it and under its shade are birds and animals. And the whole thing—tree, creatures, and all—is made entirely from decommissioned weapons: bits and pieces of old AK-47s, bullets and machetes and all the horrible paraphernalia of war, most of them made in peaceful Western countries and exported to Mozambique so that the government aid given by the West to that poor country would flow back to our own industries. The point—and it is a stunningly beautiful object at several levels—is that this particular "Tree of Life" reflects the Isaianic promise that swords will be beaten into plowshares and spears into pruning hooks. The tree stands as a reminder both of the horror of the world, with its multiple human follies and tragedies, and of the hope of new creation. It has an immediate and powerful message for the people of Mozambique, who had forgotten how to hope, had forgotten that there might be such a thing as peace, as they are invited to sit under the tree and enjoy its fruit and its healing. But it is also a sign of what genuine art can be, taking a symbol from the original creation, building into it full recognition of horrors of the present world that by themselves would lead us to despair, and celebrating the promise of the new world, a world filled with God's glory as the waters cover the sea. It offers celebration without naïveté, sorrow without cynicism, and hope without sentimentality. Standing before it is like glimpsing an apocalyptic vision, a vision of the beauty of God.

Reflecting on this vision ought to inform and direct our thinking and action in many fields of inquiry and endeavor. But for the moment we might do well simply to pause in contemplation and gratitude. This is the vision of God's new heavens and new earth; within that vision, each of us has a particular calling—prophetic, artistic, political, theological, scientific, whatever it may be—by which God will call us to bring signs of that new world to birth within the old one, where vision is still limited and widows still weep.

12

Becoming People of Hope

ONCE YOU GET the resurrection straight, everything else eventually falls into place. This point was brought home forcefully to me last year when I was in a cab stuck in a London traffic jam. The taxi driver, seeing from my clothes that I was a bishop, commented on what a difficult time we Anglicans were having over the issue of women bishops. I agreed. We were indeed having a difficult time. Then came a moment I will never forget. Turning around to face me—we were, as I say, stationary in traffic—he said, "What I always say is this: if God raised Jesus Christ from the dead, everything else is basically rock 'n' roll, i'n'it?" It was a great gospel moment, and I have dined out on it ever since.

That message is at the heart of what I want to explore. The twentieth chapter of John's Gospel is one of the most extraordinary and evocative parts of that already extraordinary and evocative book. When I was bishop of Durham, I regularly had to interview candidates for parish jobs, and I often asked them which two chapters of the Bible they would take with them to a desert island, but I added, "All right—you've already got Romans 8 and John 20." Those two chapters, in very different style and mode, contain so much gospel, rich, dense, pressed down, and running over, that I sometimes think I could just read them and nothing else forever.

> On the evening of that day, the first day of the week, the doors were shut where the disciples were, for fear of the Judaeans. Jesus came and stood in the middle of them. "Peace be with you," he said. With these words, he showed them his hands and his side. Then the disciples were overjoyed when they saw the master. "Peace be with you," Jesus said to them again. "As the Father has sent me, so I'm sending you." With that, he breathed on them. "Receive the holy spirit," he said. "If you forgive anyone's sins, they are forgiven. If you retain anyone's sins, they are retained." One of the Twelve, Thomas (also known as Didymus), wasn't with them when Jesus came. So the other disciples spoke to him. "We've seen the master!" they said. "Unless I see the mark of the nails in his hands," replied Thomas, "and put my finger into the nail-marks, and put my hand into his side—I'm not going to believe!" A week later the disciples were again in the house, and Thomas was with them. The doors were shut. Jesus came and stood in the middle of them. "Peace be with you!" he said. Then he addressed Thomas. "Bring your finger here," he said, "and inspect my hands. Bring your hand here and put it into my side. Don't be faithless! Just believe!" "My Lord," replied Thomas, "and my God!" "Is it because you've seen me that you believe?" replied Jesus. "God's blessing on people who don't see, and yet believe!" Jesus did many other signs in the presence of his disciples, which aren't written in this book. But these are written so that you may believe that the Messiah, the son of God, is none

other than Jesus; and that, with this faith, you may have
life in his name. (John 20:19–31)

Right at the start of John 20 comes the note that tells us what
it's all about—though it would be easy to miss because of the
drama that's going on. It was, says John, the evening of that day,
the first day of the week. That phrase, "the first day of the week,"
began chapter 20; when John repeats himself like this, something
important is going on. His whole Gospel is framed with echoes
of Genesis 1, starting as it does with "in the beginning." Now, he
says, the old week is over. On the sixth day, the Friday, God created
humankind in his own image, and on the Friday in John's story
Pontius Pilate brings Jesus before the crowds and says, "Here's the
man!" And by the evening of Friday, Jesus has declared what God
declared at the end of Genesis 1: it is finished. It's done. As the
Father finished the work of creation, so the Son has finished the
work of redemption. Then, on the seventh day, the Saturday, God
rests, and God incarnate rests in the tomb, his work complete.
Then—then!—"On the first day of the week, very early, Mary
Magdalene came to the tomb" and found it empty, because *this is
the first day of God's new world, God's new creation.*

You see, it has been all too easy for preachers and theologians
to imagine, within our late modern culture, that the point of the
Easter stories is to provide a happy ending after the sorrow of the
previous week or to assure us that there is life after death or something like that. But what John is saying is far more powerful and
(dare I suggest) far more relevant to our church life and witness
today and tomorrow. He is insisting that Easter is the beginning
of God's new creation, and we therefore have a job to do. The
completed work of the Father in creation and the completed work
of the Son in redemption issue directly in the ongoing work of
the Spirit in mission. That is what verses 19 and on are all about.

The scene begins, after that powerful opening, with the disciples hiding behind locked doors because they are afraid that the

Judaean authorities who arrested Jesus and handed him over to be killed will also come for them. But *there are no locked doors in the kingdom of God.* Jesus came and stood in the middle of them. "Peace be with you," he said, and he showed them his hands and his side. If you were setting this to music, you would want at that point a moment of great awe and drama, because not only do the wounds in Jesus's hands and side indicate that it really is him, risen bodily from the dead; they are the signs, the marks of love, the ready evidence that having loved his own who were in the world, he had loved them to the end, had given himself utterly for them, for us, for you and me.

Perhaps this is why the resurrection of Jesus is so hard for us to believe, as it was always hard—and not just hard, actually, but impossible. We all know that death is irreversible; people in the ancient world knew that just as we do. But that's because our world is bounded by the old creation, and Easter is the beginning of God's new creation. In the same way, we find it hard to believe that we have been loved utterly and completely by the God who made the world. We know the frailty and fickleness of human love, and we find it hard, even impossible, to imagine a love that will go all the way, that will last the course. The two go together: the Easter message is the expression of that unutterable, inexhaustible love. And that is why the faith that believes the resurrection is, in the last analysis, the same as the love that opens like a flower to answer Jesus's love with a trembling love of its own. Ludwig Wittgenstein remarkably declared that "it is love that believes the resurrection." He was right. To say, as John does, that the disciples were overjoyed is putting it mildly. George Herbert reflected this when he wrote in "The Flower":

> Who would have thought my shriveled heart
> Could have recovered greenness? It was gone
> Quite underground . . .

But he who dwelt underground on that cold Sabbath has come forth once more, and with him God's powerful love is revealed in all its glory, bringing our hearts out of their own winter into the fresh spring of Easter.

And that is where we start, where we all must start, young or old, lay or ordained, in whatever tradition we stand. Without the message of faith and love that Easter provides we are nothing, but with that message the world opens up before us as a strange, unmapped new land, full of possibilities and challenges. The disciples are not to stay in the locked room. "Peace be with you," says Jesus again. "As the father has sent me, so I'm sending you." And then, with another echo of Genesis, he breathes on them and says, "Receive the holy spirit." And in those two sentences the whole mission of the church is contained. "As . . . so."

We think back in a flash through all that has gone before in the Gospel story. Jesus comes into Galilee announcing God's kingdom, healing the sick and celebrating the good news with the most unlikely people. He challenges God's people to be truly God's people, the light of the world, now that he's there to show them the way. He speaks of the coming victory that he will win through the strangest means—through his own death. He hints at the new temple that is to be built, a new temple consisting not of bricks and mortar but of himself and his followers. He does all this and much, much more. And now he scoops up everything into one phrase, "as the father has sent me," and turns it into the great commission: "so I'm sending you." The mission of the church is not to drag people into buildings or to run raffles or issue statements. Oh, I know, buildings and money and statements matter in their place. But the mission of the church is *to be for the world what Jesus was for Israel*—a mission that will send us back to the four Gospels again and again, not only to be amazed by the power and love of God but to draw down that power and love, through prayer and the Holy Spirit, so that we can be Jesus people for the world, kingdom people for the world, forgiveness people for the world. There are no locked doors in the kingdom of God, and we who are charged to go into the world with the good news must pray them open so that the message of God's unconquerable love can get in.

"If you forgive anyone's sins," says Jesus, "they are forgiven. If you retain anyone's sins, they are retained" (John 20:23). A solemn and difficult text; we might have preferred just the first half. And yet, as always in the Gospel, forgiveness is never reduced to cheap grace, to God shrugging his shoulders and saying, "Oh well, that's all right, then." Precisely because forgiveness is forgiveness and not mere tolerance, it must go with an implacable refusal to collude with sin, with violence or prejudice or spite, with pride or greed or lust, with any of the things that deface and corrupt God's good and beautiful creation. Precisely because Easter is about *new creation,* nothing that distorts or dismantles God's creation can come there. Love demands the very best for the beloved. As a parent will not rest until the last traces of illness have been removed from the child, so God will not tolerate the disease of sin within his new creation. So the message of forgiveness is that all can be left behind, everything that we know we have done to contribute to the defacing of God's world; it can be left behind at the cross where Jesus finished, as he announced, the work of redemption. Easter says, "Welcome to God's new world," and with that we are invited to taste that forgiveness for ourselves and to hold it out to the world as the great Jubilee message, the message of hope at last.

Of course people find this difficult if not impossible to believe. The story of Thomas says it all. We live in a world of Thomases, of people who want evidence, people who don't want to be taken in, people who've been hurt before and resist what they see as a cheap and easy consolation. We live in a world where many who grew up in our churches feel horribly let down, either because the people they knew there weren't living what they professed or, worse, because some of them used their position as a cloak for greed or lust or bullying, or just because the churches seemed like a stale, dusty world of shrunken, sad, and shriveled humanness when they were discovering the beauty and glory of life and literature and love and laughter. And so, like Thomas, they hear what sounds like an old fantasy message: oh yes, it's all right really, he's alive; come back to church like you used to. And they fold their arms and say, "No, I'm not going to be taken in again, I'm not going back into that stifling little world where I felt cramped and constrained. Anyway, if I allow myself to hope like that I'll only be disappointed and hurt again." There are many like that on the edges of our churches and beyond. They want something more solid to go on.

Jesus doesn't deny Thomas what he asks for. You want evidence? Very well, here it is: here are the wounds that love has borne, here are the marks of what it cost to complete the work, to finish the world's redemption. And you see, the evidence is not just intellectual evidence, though if that's what's wanted, that is there too, in plenty. The evidence is the evidence of a living Jesus, a loving Jesus, a Jesus who has done everything Thomas needed. "Bring your finger here," he says, "and inspect my hands. Bring your hand here and put it into my side. Don't be faithless! Just believe!"

The question for us, as we learn again and again the lessons of hope for ourselves, is *how we can be for the world what Jesus was for Thomas:* how we can show to the world the signs of love, how we can reach out our hands in love, wounded though they will be if the love has been true, how we can invite those whose hearts have grown shrunken and shriveled with sorrow and disbelief to come and see what love has done, what love is doing, in our communities, our neighborhoods: the works of justice and beauty that speak of God's new creation, the works of healing and new life that should abound in our hospices and detention centers, our schools and our countryside. It is when the church is out there making all that happen, not waiting for permission or encouragement but simply doing what Christian people from the very beginning have always done, that suddenly resurrection makes sense, because suddenly the idea of God's love in new creation makes sense, and people who were formerly skeptical find their hearts and minds transformed so that they say, with Thomas, "My Lord and my God." Yes, it might have been better if they had believed without seeing. But Jesus isn't fussy. Jesus will meet Thomas halfway, because the new life of the Easter gospel is always going out to meet people halfway, to surprise them once again with the overflowing and powerful love of God. And our task, in the power of the Spirit, is always to be meeting the world halfway with the surprising signs of God's generosity. And those signs will be all the more powerful if we do them together.

One great project in which I worked with others like that happened a few years ago in the northeast of England. I worked there as bishop, and for a week thousands of young people came together to be taught the scriptures in the morning, work on social projects in the afternoon, and hold celebratory and evangelistic rallies in the evening. Among my favorite memories of the week was going with one of the dozens of afternoon groups to paint the back walls of a lane of dark and dismal houses in the wrong part of one of our old towns, and hang flower baskets all the way down the road. People were coming out of their houses—and that's something they didn't normally do, because they were afraid of those dark back alleys because of what used to go on there—and asking nervously whether we were from the council or whether they were going to have to pay. No, replied the cheerful teenagers, we're from the church; this is just a present to you. They were astonished. And the story doesn't stop there, because when I went back a year later, the residents had begun to do more things in that back alley, planting little gardens and holding barbecues and getting to know one another. One small gesture of love and generosity from the church cascaded into new life and new possibilities for a whole street. And when the lay church worker who had gone to live on that street spoke about Jesus, they knew it was true. They had seen the marks.

The story of Thomas, the focal point of this passage, is one of three astonishing personal encounters with the risen Jesus that John records. I want, in closing, just to say something briefly about the other two, Mary Magdalene and then Peter, because both of them are I think directly relevant to where we are as a church, perhaps especially when the church has gone through dark and difficult times.

First there is the story of Mary Magdalene. She is the one who arrives first at the tomb, is first to run and tell the others, first to weep in despair, first to see the angels, and above all first to see Jesus. I have sometimes wondered, reading Mary's story in John 20, whether the point is not that those who see angels are likely to do so through tears. Certainly it is often when people are at that point that the risen Jesus comes to them and surprises them with his presence and love. And then something truly remarkable happens. Up to this point (though, granted, some manuscripts vary) John has referred to Mary by the Greek form of her name, *Maria.* That's the name by which she would have been known on the street. That's what the soldiers at the foot of the cross would have called her. But when she comes to Jesus, not knowing that it is Jesus, and pours out her grief before him, he addresses her with her real name, her ancient biblical name, the Aramaic or Hebrew form, *Mariam,* "Miriam": his mother's name, the name her dad used to call her when she was little, the name that says, Miriam, it's going to be all right, I'm alive, and you are now my messenger. And Miriam, in an explosive and revolutionary moment, is the first person to tell anyone else that Jesus is alive, carrying the message at the heart of all Christian ministry. She is the apostle to the apostles.

Miriam's story is powerful, and there may be some who resonate with it at particular and personal levels. But the other story is that of Peter, in John 21. Peter, as we know, has let Jesus down badly. Three times he has denied knowing him, there beside the charcoal fire in the high priest's courtyard. Now there is another charcoal fire burning by the sea, with the smell no doubt reminding Peter, as John's reference to it reminds us, of that horrible night not long before. Jesus takes Peter apart from the others and asks him the question, the key question, the question of Easter: "Simon, son of John, do you love me?"

Peter's answer, in the Greek, uses a different word, and I cannot believe that this is accidental. He doesn't use the same word, *agápo,* which carries the full meaning of God's love. He uses the word *phileo,* a good word but farther down the scale. "Yes, Master," he says, "You know I'm your friend." That's as far as he can get. How can he say, after all that's happened, that he loves Jesus with the love he knows Jesus has shown for him? But he won't deny again, and when Jesus asks the same question a second time, he gives the same answer. "Yes, Master. You know I'm your friend."

I suspect this resonates strongly with many of us. Many of us know perfectly well that we've let the Master down badly. We have made a mess of our discipleship; we have dropped the ball, we've looked the other way, we've gotten cold feet and denied what we know we believe. And we are ashamed. Yet here we are, hanging in there, because we'd rather be near Jesus, even though we feel uncomfortable, than hiding away. "Yes, Master. You know I'm your friend." And Jesus's response is astonishing: "Feed my lambs," he says. "Look after my sheep." He doesn't say, "Well, Peter, you've really messed up; the only way forward now is a six-month penitential rehabilitation course and then we'll think about it." The word of forgiveness comes in the form of a fresh commission. That is the sign of Jesus's love. And this, remember, is Peter. All Petrine ministry begins at this point, with the free forgiveness of those who have failed.

But then comes the third question. This time Jesus uses the word Peter had used. "Simon, son of John," he says, "are you my friend?" Peter was upset, says John, that this third time Jesus said it like that, and he replied eagerly. But I think the point is this—and it's a point that I suspect many of us need, in our varied discipleship and service. This is, for me, the heart of the Easter message as applied to those who would seek to be Jesus's friends, to follow him and serve him. "Very well," Jesus is saying. "If that's where you are, that's where we'll start." As with Mary and her tears, as with Thomas and his skepticism, Jesus comes halfway to meet Peter. He doesn't insist on Peter being able to say the big *L* word, the *agápē* word, right off. That will come. Peter is hanging in there: "Yes, Master; You know I'm your friend." "All right, Peter; that's where we'll begin. Feed my sheep. And, by the way, things are going to be tough; other people will have other tasks to do, but you must simply remember this: 'Follow me!'"

It is love that believes the resurrection. It is, conversely, the resurrection of Jesus that awakens love—love for him, love for one another, love for God's world. This is the message of Easter. This is the message of hope. This is the message for, and through, the whole church, through all of us together. This is the message of Jesus. May it be so for us, in us and not least through us.

ACKNOWLEDGMENTS

A s I said in the Preface, I am very grateful to those whose kind invitations set the scene for these papers and addresses, and whose warm welcome and hospitality made the various occasions a delight.

Healing the Divide Between Science and Religion
    BioLogos Foundation Meeting, New York, March 2012

Do We Need a Historical Adam?
    BioLogos Foundation Meeting, New York, March 2013

Can a Scientist Believe in the Resurrection?
    The James Gregory Lecture 2007, St. Andrews, December 20

The Biblical Case for Ordaining Women
    A conference paper for the symposium, "Men, Women and the Church," St. John's College, Durham, September 4, 2004

Jesus is Coming—Plant a Tree!
    Seminar at Windsor Castle: "Evangelicals and the Care of Creation," July 4, 2006

9/11, Tsunamis, and the New Problem of Evil
    Church Leaders' Forum, Seattle Pacific University, May 18–19, 2005

How the Bible Reads the Modern World
    A lecture for InterVarsity, University of Chicago, November 13, 2012

Idolatry 2.0
    A lecture for the Veritas Forum, Northwestern University, Chicago, November 12, 2012

Our Politics Are Too Small
    A public lecture at the annual meeting of the Society of Biblical Literature, San Diego, California, November 18, 2007

How to Engage Tomorrow's World
    CCCU Conference, Washington D.C., February 2012

Apocalypse and the Beauty of God
    A sermon at Harvard Memorial Chapel, Cambridge, Massachusetts, October 22, 2006

Becoming People of Hope
    A sermon at the Ecumenical Service, Belfast, Northern Ireland, October 21, 2011

SCRIPTURE INDEX